自主创新 方法先行

本书的出版受科技部创新方法项目（2008IM020200）支持

当代科学文化前沿丛书

SCIENCE AND HISTORY
A CHEMIST'S APPRAISAL

科学革命新史观讲演录

〔美〕艾伦·G. 狄博斯 ◎著
任定成 周雁翎 ◎译
任定成 ◎校

北京大学出版社
PEKING UNIVERSITY PRESS

北京市版权局著作权合同登记号　01-2011-5723

图书在版编目(CIP)数据

科学革命新史观讲演录/(美)狄博斯(Debus,A. G.)著;任定成,周雁翎译;任定成校.—北京:北京大学出版社,2011.11

(当代科学文化前沿丛书)

ISBN 978-7-301-19681-6

Ⅰ.①科…　Ⅱ.①狄…②任…③周…④任…　Ⅲ.①自然科学史—世界　Ⅳ.①N091

中国版本图书馆 CIP 数据核字(2011)第 225658 号

Allen G. Debus
SCIENCE AND HISTORY
A Chemist's Appraisal
Servico de Documentação e Publicações da Universidade de Coimbra 1984
(根据科英布拉大学出版社 1984 年版译出)
科英布拉大学前校长古维亚(António Jorge Andrade de Gouveia)
和本书作者狄博斯(Allen G. Debus)授权任定成汉译和出版本书

书　　名:科学革命新史观讲演录
著作责任者:[美]狄博斯　著　任定成　周雁翎　译　任定成　校
丛 书 策 划:周雁翎
责 任 编 辑:周志刚
标 准 书 号:ISBN 978-7-301-19681-6/N・0043
出 版 发 行:北京大学出版社
地　　址:北京市海淀区成府路 205 号　100871
网　　站:http://www.jycb.org　http://www.pup.cn
电 子 信 箱:zyl@pup.pku.edu.cn
电　　话:邮购部 62752015　发行部 62750672　编辑部 62767346
出版部 62754962
印　刷　者:三河市博文印刷厂
经　销　者:新华书店
650 毫米×980 毫米　16 开本　15.75 印张　158 千字
2011 年 11 月第 1 版　2011 年 11 月第 1 次印刷
定　　价:32.00 元

谨以此书深切纪念

沃尔特·佩格尔(1898—1983)

目　　录

目　录

中文第三版前言

时间过得真快，本书汉译本繁体字版和简体字版（书名直译为“科学与历史：一个化学论者的再评价”）的出版已经过去十多年了。在这十多年的时间里，狄博斯除了继续发表研究论文之外，还于2001年出版了《化学与医学论战：从范·赫尔蒙特到波尔哈夫》（*Chemistry and Medical Debate: van Helmout to Boerhaave*, Science History Publications），于2006年出版了《化学论承诺：1550—1800年化学论哲学中的实验与神秘主义》（*The Chemical Promise: Experiment And Mysticism in the Chemical Philosophy, 1550—1800: Selected Essays of Allen G. Debus*, Science History Publications）。2002年，他的《化学论哲学：16和17世纪的帕拉塞尔苏斯科学与医学》出了第二版。

这十多年里，我的学术兴趣逐渐向科学史领域转移，但没有再跟狄博斯教授联系。朱晶博士去年告诉我，她在*Ambix*上读到讣告说，狄博斯已于2009年3月6日去世。对他的逝世，我没有发表过悼念文字。本书汉译本第三版的出版，也算是我和本书的其他译者对作者的纪念吧。

本书是一本科学史学史著作，同时也是按照作者自己的历史观对科学革命做出的新诠释。在这次汉译本第三版中，周雁翎博士建议我不要采用原来直译的书名，我就按照作者“高度个人”的观点，把书名意译为“科学革命新史观讲演录”了。另外，我对第二版中留下的少量欠妥之处又做了些处理。这是第三版的主要变化。

这本小册子是想进入科学史领域的年轻人了解科学史学科的诞生和发展、了解科学史研究进路及其变化、了解科学革命时期的复杂图景的优秀读物。作者讲的是从科学史职业化开始到20世纪70年代末科学史及其研究进路的变化，例子都是西方科学史方面的，对他本人长期研究的化学论在欧洲科学革命时期的状况有通俗简明的介绍。席文教授的《科学史方法论讲演录》也从科学史的职业化开始讲起，但主要是对20世纪70年代以来科学史研究进路的分析，中国科学史的例子较多，对作者本人及其弟子在中国传统科学和医学方面的研究工作有明晰的介绍。这两本书共同描绘了一幅较为完整的科学史学科发展变化图景，也概括地介绍了两位科学史家自己的科学史观和科学史方法论思想，可以结合起来阅读。

刘鹤玲教授和朱传方教授提供了部分章节的初译稿，为本书中文第一、二版的出版作出了贡献。谨此致谢。

任定成
2011年4月8日
于昆明至北京飞行途中

简体字版前言

本书的中文译本，1999 年 11 月由台北的桂冠图书股份有限公司出过繁体字版。当初，译完这本书后觉得应当向中文版读者作些说明的话，在桂冠版前言中都已经说过了。但是，一些见到繁体字版本的朋友，仍觉得那个前言还是没有完全告诉他们阅读这本书之前想要知道的事情。现在借出简体字版的机会，再补充说几句。

首先要说明的是，狄博斯最重要的学术贡献是什么。可以说，是他描绘了一幅不同于按照传统科学史观描绘的科学革命图景。现代科学作为一个完整的文化形态，产生于文艺复兴时期。因此，人们也把这个时期现代科学的建构过程称为科学革命。传统的科学史观认为，科学革命就是以牛顿力学为主要成就的机械论自然观战胜亚里士多德自然观的过程。狄博斯通过大量的考证和分析，发现不能简单地把科学革命看成上述两支力量的对抗，因为当时还有第三支重要力量的存在。这第三支力量就是化学论自然观及其实践者。化学论派一方面激烈抨击当时在大学里占统治地位的亚里士多德传统，另一方面又坚决反对同样激烈抨击亚里士多德传统的机械论派。当时，化学论者和机械论者都是新科学的鼓吹者。而且，在科学革命的前期，化学论的影响远远超过了机械论。

那么，化学论与机械论的主要区别是什么呢？简单地讲，在看待自然的方式上，机械论把宇宙看成是一架机器，认为世界上的一切完全决定性地按照机械运动的规律运动；化学论则把宇宙看成是一个坩埚，认为世界上的一切无一例外地都在经受化学变化。在认识结果的评价和检验方式上，机械论强调主体际性，就是要求认识主体相互之间能够重复论证或者观察到相同的结果；化学论则强调个体感悟，就是要求认识主体的主观感觉一致。

认真研究起来，化学论是一种生成论或者有机论的自然观，它在很多方面与人们所说的中国传统自然观相似。当然，中国传统自然观也不是一种。泛泛地说中国传统自然观如何如何，没有多大意思。同样，西方传统自然观也不是一种，泛泛地议论也不很妥当。但是，从狄博斯的工作，我们可以认识到，我们原来以为只有我们的祖宗才有的自然观，在西方人的祖宗那里也有了。所以说，我们以前把环境污染、能源危机、生态恶化等问题归结为现代科学尤其是“西方”机械论自然观的恶果，把“天人合一”等中国传统自然观看成是拯救人类的唯一良方，实在是有些轻率了。

抛开这些中西之争不说，还可以想到的一个问题是：狄博斯的工作是不是为中国近些年来社会上流行的在所谓中国传统文化的旗号下招摇的伪科学和反科学歪理提供了科学史论据呢？任何一位严肃的读者都会说：不是。且不说我们看待一个学说，要把它放到特定的历史境况之中，就是抛开历史背景来看，中国传统文化旗号之下的“超科学”也是不能与化学论相提并论的。化学论虽然强调个体的体悟，但是它与机械论一样，也强调观察和实验而不是信仰。单从这一点上看，化学论这种大约5个世纪前流行的学说，也是中国当代的新迷信[①]不能比的。

本书是一部科学史学史著作，主要反映的是作者的科学史学思想。希望进一步了解狄博斯的主要科学史工作的读者，还可以阅读其《文艺复兴时期的人与自然》新译本[②]。

任定成
2000年2月4日
于承泽园

---

① 对迷信和科学之间的区别感兴趣的读者可参看笔者拙文：“现代科学与新迷信之间的十个界线”，《科技日报》1999年8月14日理论版；“远离伪科学必须走出九个认识误区”，《前线》1999年第11期。

② 艾伦·G. 狄博斯著，周雁翎译，《文艺复兴时期的人与自然》，上海：复旦大学出版社，2000年。

繁体字版前言

本书篇幅不大，雅俗共赏，似乎读懂甚易。其实，由于我们对科学史领域的国际学术趋势较缺乏了解，在接触到某些科学史大师的著作时，常常对其意义认识不足。因此，在这里对狄博斯其人及其成就略作介绍。

本书作者狄博斯(Allen G. Debus)，1926 年 8 月 16 日生于美国伊利诺伊州的芝加哥市，1947 年获西北大学化学专业理学士学位，而后在印第安纳大学研读化学和历史方面的研究生课程，1949 年获得文学硕士学位。他以研究化学家的身份在阿博特实验室从事了数年制药学研究之后，于 1956 年转而进入哈佛大学研究科学史，1961 年获得这个专业的哲学博士学位，毕业后在芝加哥大学任教至今。狄博斯起初任副教授。1963 年，他在该校历史系创设了科学史与医学史研究生项目。1970 年，美国医学协会前会长莫里斯・菲什拜因(Morris Fishbein，1889—1976)及其妻子安娜(Anna)捐款，在芝加哥大学建立莫里斯・菲什拜因科学史与医学史研究中心，狄博斯即任此中心的首任主任，直到 1977 年。从 1977 年起，他任该校莫里斯・菲什拜因科学史与医学史教授。现在，他是该校的终身荣誉教授。可以说，狄博斯是芝加哥大学科学史与医学史教育和研究事业的奠基人。

狄博斯在国际科学史学界享有盛誉。他曾当选为国际科学史研究院和国际医学史研究院的院士。他还是权威的《科学史》杂志的22位顾问编委之一，是国际性的炼金术史与化学史学会的10位常务理事之一，而且经常在各种国际性的科学史学术会议上担任主席。他于1994年获得国际科学史界最高奖萨顿奖章①，1978年获得国际科学史界最高著作奖菲泽奖金，1987年获得国际化学史界最高奖德克斯特奖②。

狄博斯教授在自然科学(化学)和对自然科学的人文主义理解(科学史)两方面均有成就。在化学方面，他拥有许多专利。在科学史方面，他除发表了200余篇论文和评论外，还著有《英国的帕塞尔苏斯信徒》(《瓦茨科学史文库》之一种，1965，1966)、《17世纪的炼金术与化学》(与罗伯特·P.马尔特霍夫合著，1966)、《文艺复兴时期的化学论梦想》(1968年初版，1972年第2版)、《17世纪的科学与教育》(与布赖恩·拉斯特合著，1970)、《化学论哲学：16和17世纪的帕拉塞尔苏斯科学与医学》(2卷，1977)、《文艺复兴时期的人与自然》(《剑桥科学史丛书》之一种，1978年初版，1994年第14版；意大利文版，1982；西班牙文版，1985，1986；日文版，1986；中文版，2000)、《罗伯特·弗拉德及其哲学之钥》(1979)、《科学与

---

① 布赖恩·W.奥格尔维采访，任定成译，“1994年萨顿奖得主艾伦·狄博斯访谈录”，《科学技术与辩证法》1996年第3期。

② 任定成，“化学史成就的崇高奖赏——德克斯特奖”，《化学通报》，1992年第1期；“杰出的化学史大师——历届德克斯特奖获得者”，《科学》，第43卷(1991年)第1、2、4期。

历史：一个化学论者的评价》(1984)、《1500—1700 年的化学、炼金术与新哲学》(1986)以及《法国的帕塞尔苏斯信徒：近代早期法国化学论对医学和科学传统的挑战》(1991)等 10 部著作，主编了《世界科学名人录》(书中包括了 30000 名科学家，其中不少人仍在世，1968)，并编有《文艺复兴时期的科学、医学与社会：沃尔特·佩格尔纪念文集》(2 卷，1972)和《赫耳墨斯主义与文艺复兴》(与英格里德·默克尔合编，1988)两部文集。从与狄博斯教授的通信中得知，他仍在不停地撰写论文和著作。我们期待着阅读他的新作品。

本世纪，科学史领域取得了许多重要成果。正如尼古拉斯·H. 克拉利 (Nicholas H. Clulee) 所说，科学史和科学编史学方面的任何重大变化都不是一个人的工作所能引起的，但是狄博斯是一个例外，他“在我们的历史中给有机自然的科学赋予了一个与物理自然的数理和机械论科学并列的重要地位”，从而“剧烈地改变了我们的科学革命观以及……70 年代以来的科学史观”。①

狄博斯的许多研究工作，都是围绕文艺复兴时期的科学革命展开的。实际上这也是其他许多科学史学家们关注的焦点，因为正是在这场革命之中才形成了真正的近现代意义上的科学，而且，我们今天关于科学的本质以及科学与伪科学之间的界线的看法，在很大程度上也是由之决定的。由于受实证主义和辉格史观的长期影响，人们逐步形成了关于科学革命的传统看法。在实证主义者心目中，这场革命是进步与落

① 见 *Isis*, Vol. 86(1995), No. 2, pp. 284—285。

后、理性与非理性、机械论自然观与神学自然观、科学与伪科学、"今人"与"古人"之间的斗争。当时探索自然奥秘的人，一类是科学理性的代表，一类是保守分子或江湖骗子，而不能划归这两类集合的则是新旧思想兼而有之的古怪人。按这种传统观念描绘的科学革命，就是一幅导致机械论哲学取得成功、伽利略方法深入人心、牛顿力学得以完成的图景。遗憾的是，这幅图景很可能是不真实的，因为它是人们戴着有色眼镜，用今天的科学标准选择材料，编织历史，所勾勒出来的。

英国学者拉维茨(J. R. Ravetz)认为，佩格尔(W. Pagel)、狄博斯和拉坦西(P. M. Rattansi)是一个新的思想流派的领袖人物。[①] 我们将其称为 PDR 学派。这个学派强调要用史境进路(contextual approach)，而不是实证主义态度，去研究问题。他们深入到范围广泛的历史材料之中，发现历史的本来面貌要比用直线透视式方法所显示出来的图景复杂得多。他们认为，在一定的历史时期，在某些科学家身上，科学思想与非科学思想不是简单并列或者彼此无关地存在着的，而是一个相互支持、相互确证的有机整体。佩格尔把注意力放在哈维、帕拉塞尔苏斯和范·赫尔蒙特这三个人物身上，拉坦西主要考察各门科学与当代社会的关系，狄博斯则力图提供一幅较为完整的科学革命图景。

狄博斯发现，科学革命时期最热烈争论的问题与医学和化学有关，与化学论哲学有关。这里，"化学论哲学"(the

① J. R. Ravetz, "Respecting Aberrant Figures", *Nature*, Vol. 278, No. 5707(April, 1979), p. 813.

Chemical Philosophy)不同于我们所说的“化学哲学”(philosophy of chemistry)。后者指对化学的成就和历史进行哲学研究这样一个领域。前者主要是指从帕拉塞尔苏斯主义到赫尔蒙特主义这么一个历史时期内,带有实用化学色彩,与冶金、医学有直接联系,又具有哲学、宗教和政治蕴义的一种炼金术传统。这种哲学的真正基础建立在祈祷、信仰与想象三根支柱之上,力图用炼金术、自然法术和神秘哲学取代在大学里占统治地位的亚里士多德哲学,用化学来阐释人体和自然中的一切现象,从而达到对上帝的完美的理解。这种哲学尽管在表面上充斥着一些相互矛盾的论证,但在当时的欧洲却成为一场持续争论风暴的中心,其影响无论在烈度上还是在范围上都远远超过了天文学和运动物理学的影响。它实际上已经涉及新宇宙观的形成,新科学的建立,新实验方法的探寻,以及经济、教育和农业改革等一系列重大问题。这种哲学既反对旧的亚里士多德和盖伦传统,又与新的数理和机械论自然哲学相对抗,是早期近代科学中的一个独特组成部分,其影响曾一度超过机械论哲学及以其为哲学依托的运动物理学和天文学。因此,狄博斯的工作不仅使我们认识到科学革命的复杂性,而且唤起我们用一种新的眼光来审视世界近代史。

PDR学派的工作,实质上向传统的观念提出了严重挑战。不过,狄博斯是一位宽容的学者,他在论述科学—历史与科学史这两种传统之间的关系的文章中,就提倡分属这两大传统的学者“应当积极地彼此互助,以便在总体上促进该领域的发展”。[①] 他的风格不是锋芒毕露地经常论战，而是采用使那些恪守传统观念的人能够接受的方式去阐述自己的思想。他的作用，与其说是革命，不如说是潜移默化的影响。经验证明，他的方式是成功的，因为他的思想和成果已经被愈来愈多的人所接受。

1991 年 10 月 30 日，一些科学史学家集会芝加哥大学，纪念狄博斯教授的 65 岁寿辰，纪念他在芝加哥大学执教 30 周年，祝贺他在科学史研究中取得的成就。纪念会的主题是“体验自然”。77 岁高龄的著名科学史学家、哈佛大学教授 I. B. 科恩在会上发表了 1 个小时的基调演说。其他 11 位著名学者做了专题报告，内容涉及文化史、护身符史、炼金术史、妖术史、化学史、数学史、地学史、物理学史和生物学史，讨论的时间范围从现代早期一直到本世纪。我于会前 20 天收到了狄博斯的弟子、弗吉尼亚大学教授卡伦·帕歇尔(Karen Parshall)的会议通知，憾未克赴会，只能致函聊表贺意。[②] 但值得高兴的是，狄博斯教授的学术成就和思想已经开始引起国内学术界的关注。1987 年，《自然科学哲学问题》发表了我译出的本书第四讲的绝大部分内容。[③] 后来，陆建华等译出的《文

---

① Allen G. Debus, “The Relationship of Science-History to the History of Science”, *The Journal of Chemical Education*, Vol. 48(1971), No. 12, pp. 804—805.

② 此次会议论文后来结集作为西安大略大学科学哲学丛书第 58 卷出版。见 Paul H. Theerman and Karen Hunger Parshall (eds.), *Experiencing Nature: Proceedings of a Conference in Honor of Allen G. Debus*, Boston: Kluwer Academic Publishers, 1997。——2011 年 4 月 8 日注

③ A. G. 德布斯著，任定成摘译，“科学革命：一个化学家的再评价”，《自然科学哲学问题》，1987 年第 1 期，第 57—63 页。此译文中将作者的姓译为“德布斯”。

艺复兴时期的人与自然》纳入周谷城、田汝康二位先生主编的“世界文化丛书”出版。[①] 1987 年,《华中师范大学学报》(自然科学版)发表了狄博斯的德克斯特奖获奖演说的英文全文。[②] 1991 年,周雁翎吸收了狄博斯等人的工作,发表了从文艺复兴的广阔背景评介帕拉塞尔苏斯的文章。[③] 1992 年,刘鹤玲研究了 PDR 学派特别是狄博斯的工作对于美英科学史研究主潮的影响。[④] 此后,在一些刊物上陆续出现了有关专文或涉及狄博斯工作的文章。

狄博斯发表了不少科学史学史论文,并长期开设科学史学课程。《科学与历史》是译者所见到的唯一一部科学史学史著作,它系统阐述了作者自己的科学史学思想,同时也综合了他在哈佛大学和芝加哥大学的史学教学成果。

全书由作者在葡萄牙科英布拉大学和里斯本科学院的 4 次演讲汇集而成。第一讲“科学与历史:一个新领域的诞生”,概述了从古希腊至 20 世纪初人们对科学史的研究;第二讲“科学史:职业化与多元化”,分析了科学史成为一个职业的学术领域以来的发展状况;第三讲“化学史的意义”,讨论了化学史对于一般科学史研究的意义;第四讲“科学革命:一个

---

① 艾伦·G. 杜布斯著,陆建华、刘源译,《文艺复兴时期的人与自然》,杭州:浙江人民出版社,1988 年。此译本中将作者的姓译为“杜布斯”。

② Allen G. Debus, “Quantification and Medical Motivation: Factors in the Interpretation of Early Modern Chemistry”,《华中师范大学学报》(自然科学版),第 23 卷(1989 年)第 1 期,第 137—148 页。

③ 周雁翎,“帕拉塞尔苏斯:新科学运动的领袖与怪杰”,《自然辩证法通讯》,第 13 卷(1991 年)第 5 期,第 69—79 页。

④ 刘鹤玲,“本世纪 50 至 70 年代美英科学史研究的学术走向”,《自然辩证法研究》,第 8 卷(1992 年)第 10 期,第 28—34 页。

化学论者的再评价”,介绍了作者对于科学革命所作的新诠释。

迄今关于科学史及其历史的描述大都采取机械论立场。本书的书名告诉读者,这是一个化学论者而非机械论者对于科学与历史所作的诠释。因此,作者一再申明本书是他“关于科学史的高度个人的观点”。不过,它给读者的印象虽然似乎是在客观地讲述引人入胜的科学史学史故事,实则是在科学史学史的整体背景之中展示自己的科学史成就的意义,通过这些故事让读者历史地得出他所得出的结论。书中的巧妙之处在于,作者既对科学史各个发展时期的主要背景、特点、事件、人物和著作进行了介绍,又有意识地把自己的科学史观融进史学史叙述之中,使读者在获得科学史学史知识的同时,于不经意之中接受作者的观点,毫不反感地欣赏由他勾勒的另一幅科学革命图景。全书 4 讲均可独立成篇,放在一起又相互连贯,结构紧凑自然,口语化的文体读起来很轻松。

需要特别指出的是,书中的第三、四讲涉及的全然不是狭义的化学史,而是从一个全新的视角对科学史和科学革命所作的具有普遍意义的讨论。

鉴于国内目前尚无科学史学史的书籍可资参考,狄博斯教授 1985 年把本书寄给我以后不久,我就组织翻译出了初稿。其中大部分译稿曾打印出来。许多见到这个译稿的同行和朋友都建议我们将译稿整理出版,以填补国内科学史学史书籍出版的空白。愿本书中译本的出版,能够促进我们对于科学史学史的了解与研究。此外,由狄博斯的工作,我们可以看到，东西方文明的界线并不是传统上认为的那样分明。[1] 从这个意义上看，对狄博斯工作的重视，也许会给我国学术界和文化界近年复兴的关于中西文化、科学与人文、道与器、传统与现代、所谓"后现代"文化与社会以及所谓"文明的冲突"等问题的讨论，提供全新的启示。所以说，本书适合对文化史、思想史和科学史中的新思潮、新方法和新成就感兴趣的读者阅读。

关于本书的译述工作，有六个技术性问题需要略加说明。第一，我和我的学生们在我们以前发表的有关译文和论文中，一直按照新华通讯社译名资料组编、商务印书馆出版的《英语姓名译名手册》(1985 年修订第 3 版)，把作者的姓"Debus"译作"德布斯"。今据与 Debus 直接交往过的芝加哥大学博士生 Mr. Ryan Boynton 的发音，将其改译为"狄博斯"。第二，我们以前根据各种英—英和英—汉词典，把"the Chemical Philosophy"译为"化学的哲学"，从以上介绍可以看出，这种译法在字面上易与我们今天所说的"化学哲学"(philosophy of chemistry)相混淆。本书在将其改译为"化学论哲学"的同时，相应地根据上下文，在有关地方把"chemical"译为"化学论的"，以与"机械论的"(mechanical)相对应；把"chemist"译为"化学论者"，以与"机械论者"(mechanist)相对应。第三，"historiography"一词，既指以客观的历史实在为研究对象的活动、作品、历史、思想和方法，又指以这种活动、作品、历史、

---

① 刘鹤玲，"帕拉塞尔苏斯运动：西方文化传统中的天人合一"，《方法》，1997 年第 9 期，第 13—14 页。

思想和方法为研究对象的学科。目前中国科学史界，一般将其译为“编史学”，笔者及其学生们，以前也采取这种译法，但这种译法似难以表达原意。今采用史学界的一般译法，将其译为“史学”，以此就教大方。第四，作者注中的参考文献，凡我们知道有中译本（文）的，在译文中均相应地一一列出，以便读者查阅。第五，为便于中文版读者阅读，我在译本中加了少量注释，以脚注形式列在相应页页末。第六，原版书末附有人名索引，但所列人名有少量遗漏，我们在中译本中将遗漏部分作了补充，将人名索引改为西中人名对照表。

科英布拉大学前校长古维亚（António Jorge Andrade de Gouveia）教授，帮助译者获得中译本出版授权；狄博斯教授多次惠赠大作，并在拉丁文、西班牙文和古英文的翻译上，给译者诸多帮助；金吾伦先生审校了本书第四讲译稿。我们在此一并向他们致谢！

任定成
1998 年 1 月
于北京大学

原　序

这里刊出的，是 1983 年 4 月 27 日和 29 日，以及 5 月 2 日和 4 日，在科英布拉大学的 4 次演讲。此外，部分地以头两次演讲为基础，5 月 5 日在里斯本科学院还作了一次演讲。[1]作这几次演讲，完全由于该校前校长和化学系前主任 A. J. A. 德·古维亚（A. J. A. de Goureia）教授，他对于化学史的兴趣起初导致了关于这个学科的通信，接着便是该系和该校的联合邀请。

这些演讲就其归属而言算是史学方面的，这是由于我对这个学科持久不衰的兴趣所致。[2]任何一位在 50 年代受教育的科学史学家或医学史学家，对于自那时以来这个学科中所发生的许多变化都很清楚。即使在乔治·萨顿（George Sarton）1956 年去世之前，他的实证主义就处于抨击之中，而且今天仍然有各种相互竞争的看法、彼此不一的方法论以及大相径庭的诠释可供学者们选择。我们的学生阅读这些观点各异的论著时受到的影响，使我于 15 年前在芝加哥大学为我们科学史专业一年级的研究生开设了一门史学必修课。我所希望的，就是向这些学生展示范围广泛的科学史与医学史论著，使他们能够在这个领域中更好地自我定向。我也希望指出以往的历史诠释如何变动不居，来教育他们对其他观点更加宽容。

事实上，我确实感到当今科学史和医学史中最大的需要是宽容。目前我们看到，当人们跨进了至少在一个世纪以前其意义还未被认识的新研究领域的时候，僵化有增无减。由于这些学者发掘出重要材料，他们便厌烦起那些仍然固守较传统的观点的人们来。但是，教授科学史和医学史的人必须明白，驱逐了他们的同类，他们也就不会生存了。相反，他们的任务，是在有重大意义的领域里捕捉潜在的学者，力图给他们以他们所需要的另外的工具，使他们能够做出独特的贡献。倘若他们已经被当作科学家来训练并且愿意继续停留在公认的内在主义传统之中，那么应当得到鼓励。倘若他们来自把科学与我们的文化和社会的其他方面联系起来的领域，也应当受到欢迎。的确，传统的做法并没有为我们提供我们这门学科的必要框架。然而，各种较新的诠释终将使那些看到科学和医学对于世界史的重要性的人，以及那些经常在科学史文献中遇到技术性论文而失却理解希望的人，都对这个领域感到满意。

我相信，科学史学史和医学史学史对于训练那些希望成为这个领域的专家的学生来说，是重要的。不过，这些演讲也可唤起人们对于一门特殊的科学即化学的历史意义，以及我本人关于化学史在科学革命[1]发展中的作用的见解的注意。科英布拉大学对我的邀请与这样一个事实紧密相连，即通过A. J. A. 德·古维亚教授、A. M. 阿莫瑞姆·达·科斯(A. M. Amorim da Costa)教授以及化学系的努力，这里最近才开设了一门科学史课程。他们的兴趣主要集中在化学史上。由于这个理由，我的这些演讲在很大程度上利用了我自己的背景和研究。

接受化学家训练时，我被科学史所吸引，因为我想对化学在科学革命时期的作用有较多的了解。然而，当我成为哈佛大学科学史专业的研究生时，立即发现化学被人们当作近代科学发展中的一个相对无足轻重的因素而被普遍忽视了。这样，我就面临着我自己的一个史学问题，即用实例论证这个被人们相对忽视了的主题应当引起学者们的注意。由于这个原因，我在那些年的研究集中在修正我的研究生时期占统治地位的亚历山大·柯瓦雷(Alexandre Koyré)的内在主义传统上。

正是由于种种原因，我决定就这些演讲的主题，首先讨论科学史作为一个领域的发展，然后特别地转向化学史及其对于理解科学革命的重要性上去。相应地，第一讲粗略地勾勒了从文艺复兴时期科学史和医学史的撰写到乔治·萨顿的工

---

① “科学革命”有广义和狭义之分。广义的“科学革命”(scientific revolution)泛指科学史上任何一次根本变革，狭义的科学革命(the Scientific Revolution)特指从文艺复兴时期开始的近代科学体系的诞生和形成。本书所说的“科学革命”多指后者。

作，指出了历史所能反映的历史学家们根深蒂固的信念的程度。第二讲讨论萨顿去世以后的时期，指出新的研究领域以及对实证主义传统的挑战。第三讲转入化学史，指出其发展何以不同于一般科学史的发展。最后一讲主要介绍我自己较早期的研究，说明 16 和 17 世纪化学的发展确实是科学革命的一个基本部分。

这些评论可能已经使读者得出结论，认为这些演讲所介绍的完全是一种个人的观点。这当然是对的。我并非认为不是如此。我没有使它们实际上成为综合性的演讲的打算，而且肯定有许多人宁可见到讨论其他作者的著作而不是我所选择的著作。我也确信许多读者对这个领域的发展所描绘的画面与我所描绘的会完全不同。我期待着如此而且确实希望如此，因为这是从这个领域的文献中所要吸取的教训之一。毕竟，我们反映的是我们自己的背景，而且，在科学史这样一个多变的领域中，不存在两个受过相同的科学、哲学和历史训练的人。让我补充一个防止误解的最后的说明：读者将会发现，书中提到的作者重点是英美人。我不为此道歉。萨顿首先影响的便是美国的科学史的发展。正是在美国和英国，我接受了训练并且开展了我的大部分研究。此外，这些作者是最为我们的学生所理解的人。

最后，我要感谢德·古维亚教授、达·柯斯塔教授以及他们的家庭在我和我的妻子在葡萄牙期间对我们的款待。感谢科英布拉大学校长罗·阿拉科(Rui Alarcão)教授、副校长乔治·S. 维伽(Jorge S. Veiga)教授、化学系主任约斯·西蒙斯·雷定哈(José Simões Redinha)教授、该系全体成员以及惠予出席这些演讲的其他人。也感谢里斯本科学院院长约斯·宾托·佩克苏托(José Pinto Peixoto)教授友好的接待,感谢我们在该地会见的科学院的许多成员。确实,正是由于我们所见到的与这个国家一样美好的人民,我们对葡萄牙的访问才会是最值得我们永远纪念的经历。在某种意义上,我受惠于我的史学课程中我与学生们多次活跃的讨论,过去15年他们反对这里所描述的内容。我还必须感谢北伊利诺伊大学的乔·D. 本奇菲尔德(Joe D. Burchfield)教授阅读并评论了前三讲初稿,感谢威斯康星—麦迪逊大学的约翰·纽(John Neu)帮我弄到了在芝加哥大学弄不到的几部珍本书。

1976年《科学年刊》(*Annals of Science*)编辑I. 格拉顿—盖英内斯(I. Grattan-Guinness)博士约我为他计划在该刊上发表的"关于科学史的个人观点"系列文章准备一篇论文。我的其他许诺使得撰写此稿的任务落空了,不过我确实为准备这篇论文撰写了一些札记,而且万一他碰巧读到这些内容,知道它们已经组合成这篇序言以及这些演讲,[3]他或许会高兴的。

最后我要提到沃尔特·佩格尔(Walter Pagel)。他与我之间四分之一世纪以上的友谊导致在磨碾山(Mill Hill)他家中的数百次讨论,这些讨论深深地影响了我自己关于这个学科的观点。说来奇怪,正是在为这些演讲撰写我自己关于他的工作的影响的评论之时，我接到了伦敦大学学院的 P. M. 拉坦西(P. M. Rattansi)教授的电话，他告诉我佩格尔教授于 1983 年 3 月 25 日去世。我感激他给予我的恩惠，把这些演讲献给他。

艾伦·G. 狄博斯

1983 年 9 月 14 日，美国，伊利诺伊，鹿野

注

[1] 将以"今日科学史"为题，发表于 *Memórias da Academia das Ciências de Lisboa*，并载于将由里斯本科学院在葡萄牙出版的一部科学史单行本中。

[2] 这种兴趣使我在那些年里发表了许多涉及科学史学史的论文。其中可列举的有："An Elizabethan History of Medical Chemistry", *Annals of Science* 18 (1962, 出版于 1964 年), 1-29; "The Significance of Early Chemistry", *Cahiers d'Histoire Mondiale* 9 (1965), 39-58; "Alchemy and the Historian of Science", *History of Science* 6 (1967), 128-38 [乔斯顿(C. H. Josten)的 Elias Ashmole 的评论文章]; "The Chemical Philosophy of the Renaissance", in *The Rise of Modern Science: Internal or External Factors?*, George Basalla ed. (Lexington, Mass: D. C. Heath, 1968), pp. 82-88[这是一部从惠威尔(Whewell)到目前

有关科学革命的不同诠释的有用的选集。它可能会重印]；"Chemistry and Scientific Revolution", in *Teaching the History of Chemistry. A Symposium*, San Francisco, California U. S. A., April, 1968 (Budapest: Akadémiai Kiadó, 1971), pp. 101-11；"The History and the History of Science", *Ambix* 18 (1971), 169-77；"The Relationship of Science-History and the History of Science", *The Journal of Chemical Education* 48 (1971), 804-5；"Some Comments on the Contemporary Helmotian Renaissance", *Ambix* 19 (1972), 145-50 [重印J. B. 范·赫尔蒙特(J. B. van Helmont)的 Aufgang der ArtzneyKunst (1683；1971)的评论文章]；"The Pseudo-Science and the History of Science", *The University of Chicago Library Society Bulletin* 3, (1978), 3-20；"The Arabic Tradition in the Medical Chemistry of the Scientific Revolution", in *Proceedings of the First International Symposium for the History of Arabic Science, April* 5-12, 1976, Ahmad Y. Al-Hassan, Ghada Karmi and Nizar Namnun eds. (Aleppo: Institute for the History of Arabic Science, 1978), 2, pp. 275-90；"The Geber Tradition in Western Alchemy and Chemistry", in *Proceedings of the Second International Symposium for the History of Arabic Science*, 5-12 *April* 1979 (Aleppo: The Journal of the History of Arabic Science, 印刷中)；Obituary of Walter Pagel, *Bulletin of the History of Medicine* (印刷中)。

[3] 西里尔·斯坦利·史密斯(Cyril Stanley Smith)为格拉顿—盖英内斯的系列文章撰写的论文的标题为"关于科学及其历史的高度个人的观点"[Annals of Science 34 (1977), 49-56]。这倒很适合用作我本人在科英布拉大学所作的系列演讲的标题。

Chapter 1

第一讲

科学与历史：一个新领域的诞生

由于迄本世纪中期为止萨顿几十年遍及世界的活动，此领域反映出他的偏见也就不怎么奇怪了。第二次世界大战后期，少数哲学博士从哈佛大学和其他大学的这个新领域毕业了，但这些年轻学者的著作以及《爱西斯》上的文章仍然反映出强调物理科学发展的实证主义倾向。这种情况以后要根本改变。

35年前，当我还是西北大学的学生时，科学史还没有作为一个研究领域而存在。我的主修专业是化学，当时该系确实提供了一门不长的历史讲座课程，但知道这件事的学生寥寥无几，而听课的人则为数更少。当时，我兼修了一门历史课，但该系的有关成员则认为这个科目要取消。这门课固然承认有过那么一场科学革命，但却从未告诉我们它对于世界史的意义，也未涉及过去100年的科学和技术。标准的教学内容强调的仍然是政治、社会和宗教。甚至就连我们听到的有关智识史的极少几次演讲，似乎也对科学漠然视之。历史专业的学生的印象是，世界史并未受到科学和医学兴起的影响；而对于主修任何一门科学的学生来说，世界史是不应理会的学科，因为他们担心被陈腐的理论和事实把头脑搞得杂乱如麻。

当然，现在我明白，我当学生时的印象错了。科学史与医学史的研究是重要的。此外，尽管当时我不了解，但现在却知道科学史与医学史的撰写已经有悠久历史。甚至在古代也是如此。众所周知，普罗克拉斯(Proclus)和塞尔苏斯(Celsus)就撰写过几何学和医学简史。[1]这一传统在中世纪亦未中断。居伊·德·肖利亚克(Gui de Chauliac)为其著名的《外科学》(*Surgery*, 1343)所作的序，就是一篇对这个领域的历史概述。[2]这些早期的历史著作确然不多，但却足以表明，当时人们就已意识到，不仅知道自己领域的学科内容重要，而且了解其历史发展也很重要。

详细讨论科学史和医学史的撰写是一件相当重要的事，但在我所拥有的短时间内，没有足够的时间去做这件事。今天我所讨论的范围主要是16世纪晚期至20世纪中期近400年的时间，并且仅提出几个论点。我想指出，17世纪中期前后是科学的大分水岭，人们撰写的科学史范本存在着差异。这些作者的科学信念的确影响了他们的历史观。我还想进一步论及启蒙运动和19世纪早期的科学史，指出它们延续到本世纪的影响。

文艺复兴时期的科学史：帕拉塞尔苏斯信徒与古人

我们首先转入对科学革命的讨论。正是在16和17世纪，我们才第一次看到数目甚多且不断增长的科学方面的历史著作。这些著作向我们展示了各个时期的历史著作的一种最显著的特征，即历史学家们总是心怀某种目的去撰写历史。实际上，不论科学史学家自觉还是不自觉，他们通常都是宣传者。20世纪60年代后期，我们目睹了由激进的历史学家们组成的一个新学派的诞生，他们的工作顺应了时代倾向。在16和17世纪，我们看到了类似的情况。当时的宗教和科学中的改革运动反映在当代历史学家们的著作中。这样，帕拉塞尔苏斯(Paracelsus，1493—1541)的医学和化学改革可以在其信徒们的著作中找到。[3]这些改革实际上既有理论上的也有实践上的。一方面，他们寻求一种以神秘宇宙观为基础的对世界的新理解，以通过人去诠释微观世界以及周围的整个宏观世界。人的小世界中的一切都可以在宏观大世界中找到。而人作为真正的自然法术①师，在对上帝所创造的自然界的研究中，则可以听到造物主的声音。化学可能是这种新知识的关键，因为自然和人均可通过化学过程及类似过程得到最好的理解。因此，如果帕拉塞尔苏斯信徒为赞成知识的某个新理论基础而辩论，那么，他们还可看到一种对实践改革的需要，因为人体生理学是用化学或炼金术术语描述的。用化学方法制备的新药品用于遏制那些被认为是体内的化学失调现象。他们认为这些化学药物远比盖伦草药合剂有用得多。简言之，16世纪帕拉塞尔苏斯的医学和科学是反盖伦学说和亚里士多德学说的。它倡导新的观察，并论证为了医生的利益，这些观察完全只能由化学论者们进行诠释。

这一要旨被许多用历史支持自己信条的作者作了详尽阐述。让我通过16世纪两位作者的著作来说明这一点。第一位主要是对帕拉塞尔苏斯哲学的理论方面感兴趣，而第二位则主要把自己限定在采用化学方法制备新药品的实践方面。前者的著作的标题很长，是该时期的典型写照：《由圣先首先传授的由统一性、和睦与协调所组成的古代医

① 自然法术，即不借助神力的法术。

学，与来自盖伦之类的偶像崇拜者、异族人和异教徒的由二元性、倾轧和对立所组成的后来的医学之间的差异》(*The difference betwene the auncient Phisicke, first taught by the godly forefathers, consisting in vnitie peace and concord: and the latter Phisicke proceeding from Idolaters, Ethnickes, and Heathen: as Gallen, and such other consisting in duality, discorde, and contrarietie*, 1585)，作者是绅士 R. 博斯托克(R. Bostocke)，他抨击各大学正在讲授的错误的自然哲学和医学。尽管经验已经能够证明化学药品的效力，但学生如何知道呢？

> 在各个学校里，那些非出自亚里士多德(Aristotle)、盖伦(Gallen)、阿维森纳(Auicen)以及其他异族人的任何东西均不能得到承认或认可，也不能让那些年轻的初学者们知道这种学说，不然就让他们憎恨这种学说。同样，在国外，盖伦主义者们被王室的保护、特权和权力如此武装起来，如此保护起来，使得不被他们所允许的任何东西都不能得到允许，使得不同意他们的意愿和学说的任何东西都不能得到承认。[4]

对于博斯托克来说，正如对于这一时期的其他作者一样，宗教是一个主要因素。亚里士多德和盖伦是异教徒。他们的错误的哲学和医学是被那些未经确证就去解读和评价其著作的讲师们弄得久存不衰的。追求真理的人应当向上帝，而不是向古人的著作学习。“天国和尘世的全能造物主(基督教读本)，在我们面前展示了两部最重要的著作，一部是大自然，另一部则是《圣经》。”[5]

这是一种不同的研究自然的方法，它提倡摧毁古人的哲学，并由基于《圣经》、观察和实验的新科学取而代之。这大概是一种基督教哲学。

对于博斯托克来说，历史是一种重要工具。他几乎有一半的论著是关于化学史和医学史的。[6]由于相信可以从已被人们认为是几乎陈旧了的《旧约全书》和《人体奥秘》(*Corpus Hermeticum*)中恢复传授给亚当(Adam)的原始知识的原状，因而他论证说，这些真理已被部分地保留在前苏格拉底信徒们和柏拉图(Plato)的著作中。但是，亚里士多德抨击了他的老师，而采纳亚里士多德哲学的盖伦则通过迫害基督徒对他的罪孽作了妥协。在以后的几个世纪中，对于大部分拜占庭和伊斯兰炼金术士来说，只有少数几个有贡献的人保留了最古老的真理，将其由师傅到徒弟地传授下来。这样看来，帕拉塞尔苏斯就不是一个革新者。当然，他的医学改革完全可与重新发现了真正的古代天文学的哥白尼(Copernicus)的改革以及重新发现了古代神学真理的路德(Luther)、梅兰希顿(Melanchthon)、茨温利(Zwingli)和加尔文(Calvin)的改革相媲美。[7]

正如所料，这样的看法遭到那些引亚里士多德和盖伦著作为学术界之荣的人们的抨击。托马斯·伊拉斯都(Thomas Erastus，1524—1583)就是这样一位作者，他把帕拉塞尔苏斯描述成一个宁可相信法术和魔鬼，也不相信权威的经典的无知庸医。然而，更有趣的是那些寻求妥协的人们的反应。约翰·阿尔伯特斯·威姆帕纽斯(John Albertus Wimpenaeus,1569)看出了帕拉塞尔苏斯工作中的价值,但他并不否定古人,因为在二者中都能找到智慧。[8]安德纳赫的居恩特(Guinter of Andernach,1505—1574)甚至更有趣,他也许是文艺复兴时期最著名的医学人文主义者。[9]作为一个年轻学者,他翻译了盖伦以及埃伊纳岛的保罗(Paul of Aegina)、奥利巴苏斯(Oribasius)和查理斯的亚历克山大(Alexander of Tralles)的许多著作,而作为巴黎的一位医学教授,他又培养了安德烈·维萨留斯(Andreas Vesalius)和迈克尔·塞尔维特(Michael Servetus),此二人都曾是他的助手。

从许多方面看,居恩特都是典型的学者,因为他从没有停止研究和学习。这样,我们发现他在晚年仍认真阅读帕拉塞尔苏斯的新医学著作。但是,一个博学的盖伦主义者如何评价这些著作呢?居恩特在其1571年出版的巨著《论新旧医学》(*De medicina veteri et nova*)中作出了回答。在此书中可以看到,他提出了对于帕拉塞尔苏斯派作者来说是很基本的一个关于纯化学药品的见解。的确,他写了那篇"化学论者的医学不仅仅是神授"。[10]然而,这个人文主义者仍坚持认为医学的理论基础必须停留在盖伦主义的基础上。帕拉塞尔苏斯的思想神秘得令人讨厌,而为其辩护者亦傲慢自大。

居恩特的困惑是,既要驳斥帕拉塞尔苏斯及其狂热信徒,同时又要保持化学药品的好处。这个目的与博斯托克那样的帕拉塞尔苏斯信徒的目的截然不同，但正如博斯托克一样，居恩特为了寻找他的答案而转向研究历史。[11]他写道，最早的人身强体壮，只需简单的些许治疗就可保护其健康。较使人委靡衰弱的疾病只是后来几个世纪积聚的奢侈导致人类永久堕落的时候才产生的。正是在这时，我们看到各种不同的药品——树脂制品和芳香物质——在伊斯兰和印度学者的著作中作了介绍。帕拉塞尔苏斯的命运不仅是要恢复使用其他学者所熟知的化学品，而且还要用新的水、汁、盐、油等宝藏——通常比传统药品更灵验的药品来丰富它们。因此，居恩特对当时的医学辩论的回答就是折中。两种医学都需要。“由于时间的缘故，受尊重的权威古人应该置于首要位置”，但在较近期的化学论者的工作中，有很多具有伟大价值的东西。要是盖伦更加简洁更加精确该有多好；要是泰奥弗拉斯托斯（Theophrastus，即帕拉塞尔苏斯）更加坦率更加公正该有多好！两派的著作中糟粕精华并存，但是医生必须取二者之精华。[12]

与博斯托克相比，安德纳赫的居恩特并没有期求以恢复亚当所知道的原始医学作为手段，来证明帕拉塞尔苏斯的真理是古代就有的。相反，古代医术再次归功于希腊学者。化学药品由阿拉伯医生所采用。它们后来被人遗忘，最后又被帕拉塞尔苏斯重新发现。这已经是一项卓著的成就，他应当因此而受到称颂。但他的神秘宇宙论则可以审慎抛弃——否则就会被简单地认为是对希腊人所知概念的重述。

启蒙运动中的科学史

如果科学革命沿着帕拉塞尔苏斯信徒们指引的路线前进，那么博斯托克的炼金术史也许今天仍可读到。但事情并非如此。17世纪机械论者的成就，不是把重点放在化学和医学上，而是放在天文学和运动物理学上。这种变化亦可在同期的历史著作中看出来。17世纪晚期和18世纪的机械论者试图把他们自己从神秘主义，从简直就与他们所信赖的希腊哲学家一样的前辈们的法术中分离出来。但当他们写到"古人"与"今人"的冲突时，他们的意思尤其是指希腊哲学家的信徒们与机械论哲学家的对立。这些机械论者发现原子主义是他们追求一种以物质的很小的组成部分的尺度、形状及运动为基础的解释性模型的有用工具。数学抽象成为他们手中分析自然现象的强有力工具，而艾萨克·牛顿(Isaac Newton)的《数学原理》(*Principia mathematica*,1687)几乎成为新科学的圣经。[13]

由于牛顿对于18世纪具有重要作用，我们停下来看一看约翰·弗赖恩德(John Freind,1675—1728)的工作也许是有趣的。弗赖恩德是牛顿的门徒，占据着牛津大学的化学教席和医学教席。他在化学和医学上均有著述出版，而且正如我们所料，他的观点染上了他所信仰的科学的色彩。他的化学工作就是企图把自己与较早的化学家们区分开来的一种公开尝试。的确，弗赖恩德试图把化学反应解释为与牛顿所假设的太阳系的万有引力相类似的力所控制的球形原子间的相互作用。这是建立一种牛顿化学的一个尝试。[14]

但是，正是在1725和1726年出版的弗赖恩德的医学史中，我们才能将其观点与博斯托克以及较早的帕拉塞尔苏斯的辩护士们的观点进行比较。对于弗赖恩德来说，帕拉塞尔苏斯信徒们的神秘宗教观是不能容忍的。弗赖恩德反对帕拉塞尔苏斯，认为他是一个无用的分类者，并认为其整个宇宙论以及关于自然的宗教—生机论观点是新科学的真正对立面。另一方面，到18世纪这个开放的年代，对于用化学方法制备的药物的价值似乎已无须怀疑。然而，弗赖恩德并不愿容许人们把这些药物的发现归功于帕拉塞尔苏斯。相反，如同居恩特一样，他坚持认为应把发现这些药物的荣誉归功于阿拉伯化学家和医生。[15]……弗赖恩德和博斯托克的历史多么不同！它们都是医学史，但却是根据被17世纪中期科学分水岭分开的两种相反的观点写出来的。

对于18世纪的哲学家来说，牛顿的榜样和新科学意味着一个新时代的诞生。这就告诉历史学家要放弃其传统研究。在国王、教皇和战争的故事中，找到了什么道德价值呢？科学史是十分不同的，因为，在科学史里，我们目睹了被真正的人类英雄们从愚昧中创造的人类进步的崇高典范。后来该世纪的一位报界人士写道：

> 最近10年间各种难以置信的发现成倍地增加……人类的冒险所探索的电学现象、转化的元素、分解和理解的空气、聚集的阳光、全面考察的空气，以及数以千计的其他现象奇妙地扩展了我们的知识疆域。谁知道我们能走多远？什么凡夫俗子敢限制人类的心智呢？[16]

不仅科学会继续进步，而且，其历史也许会允许我们预测未来。正如盖尔巴特(Geolbart)所写的，

> 到本世纪末，这种对科学内在力本论的信仰变得明晰得多，并且展示了一个十分乐观的未来图景。这种曾把人类推进到现在这种科学精密和专业技术状态的同一种力仍在起作用，并将鞭策人类奔向未来。尽管并非总是稳定地进步，但这种进步将把人类带向不能想象的理智之巅。科学的昨天为未来提供了各种有价值的线索。对真理的追求并没有终止。尽管我们很伟大，但我们的科学乃至我们子孙们的科学亦将被超越。[17]

相信能够设计未来就是去证明一个尚未实现的希望，但是对作为进步根据的科学史的重要性的深信不疑，却成了18和19世纪科学的特征。纪瑟夫·普里斯特利(Joseph Priestly)的电学史和空气气体学史、J. E. 蒙丢克拉(J. E. Montucla)的数学史以及让·西尔万·巴伊(Jean Sylvain Bailly)的各种天文学史，现在仍然是学者们的研究对象。[18]正如我们在波尔哈夫(Boerhaave)的《化学新方法》

(*New Method of Chemistry*)和拉普拉斯(Laplace)的《世界体系》(*System of the World*)中所看见的那样，这是科学家们开始在其科学论著中引入他们自己的学科史的时期。[19]

然而，按这种方式，这是一部与文艺复兴时期帕拉塞尔苏斯信徒们的历史一样带有偏见的历史。这些历史学家们描写了导致现代科学形态得以形成的以往的进步。其重点总是放在西欧的科学上。中世纪的宗教气味受到蔑视，受到谴责，因为那时没有什么科学进步。他们对远东和伊斯兰的成就亦考虑甚少。

19 世纪的科学与宗教

启蒙运动中认为科学本质上是进步的这一观点给我们这个时代打上了烙印，只是在近几十年，人们才把科学史置于更广的前后历史境况中对历史进行探索。请允许我再回顾一下我当研究生时所受的训练。那时，第一部真正的科学史——与 18 世纪普遍流行的各门学科的科学史相比，这是一部综合性的科学史——被认为是威廉·惠威尔(William Whewell)的《归纳科学史》(*History of Inductive Science*,1837)。惠威尔是维多利亚时代英国一位伟大的物理学家，这个书名反映了他对培根科学思想的信奉，而培根的科学主要是一种归纳的，而不是演绎的科学——是一种以观察和实验为基础的科学。正如培根一样，他试图抛弃过分依赖数学的科学。也正如培根一样，惠威尔觉得科学史的主要目的是为科学哲学提供材料。在惠威尔看来，历史服从于哲学，或者换句话说，我们历史学家的目的是阐明科学方法。[20]

惠威尔的历史留下了众多领域尚未涉及。他忽视古代近东的成就，部分原因是他写作时可靠的资料不足，也因为他觉得埃及和巴比伦的科学缺乏理论。远东和伊斯兰成就的遭遇也好不了多少，而由于希腊科学的演绎性质他甚至对希腊科学亦进行了苛刻的描述。至于他对科学革命期间及其后的科学的描述，显而易见，受到了启蒙观点的影响。书中各章分别论述各门科学或者现代研究的各个领域。当然，此书并不是一部完整的历史，由于惠威尔个人的兴趣，书中几乎没有涉及生物科学和医学科学。

正是在惠威尔论述中世纪的章节中，他的偏见极为明显。[21]在他看来，这是由于忽视物理论证所造成的与基督教相应的“停滞期”。这一时期的所谓科学家什么新知识都没有增加。他嘲笑了他们的“不清晰性”、“教条主义”、“神秘主义”和“注解风气”。诚然，他提出有那么几个例外——罗吉·培根（Roger Bacon）和大教堂的建筑师们被他所铭记，但总的说来，他因为厌恶，因为相信这一时期物理科学只不过是法术，而拒绝考虑这一千年的历史。他带着明显的解脱心理抛弃了这个教条主义时期。

> 人类知识停滞期的惰性和盲目性的原因，终于开始抵挡不住倾向进步的原则的影响了，思想的不清晰

> 性是健全的知识衰落的基本特征。可靠的纯数学和天文学素养以及人文学科中的各种创造，唤起并强化了我们关于各种自然现象之间关系的概念的清晰性，并且对这种思想不清晰性作了几分补正。人们的思想清晰之时就是他们的奴性减少之时：对真理本性的感知把人们从纯粹观念的论战中引开；当他们清晰地看到各种事物之间的关系，就不会把全部注意力放到那些已经被别人谈论过了的问题上；因此，当科学有了见解时，注释之风也就走进了死胡同。[22]

无论在什么地方，我们只要留意一下19世纪的历史学家们，就会发现与惠威尔相似的观点。W. E. H. 莱基(W. E. H. Lecky)写下了《欧洲理性主义精神兴起和影响的历史》(*History of the Rise and Influence of the Spirit of Rationalism in Europe*, 1865)，乔治·居维叶(Georges Cuvier)给他对巴黎科学院近期工作的全面述评所冠的标题就是“1789年以来的自然科学进步史”(*Histoire des progrès des sciences naturelles depuis*, 1789)，当然奥古斯特·孔德(Auguste Comte)的实证主义以科学进步为基础。所有这些本质上渗透着启蒙精神的著作，都正视着中世纪科学和早期近代科学的强烈对比。正如惠威尔所指出的，前者与罗马天主教会统治时期相一致。另外一些人则重视科学革命发生于基督教改革运动及其直接后果之中这一事实。两种情形都迫使学者们去考察科学与宗教的关系，而且这种关系成为19世纪晚期开始争论的主题。这也许在许多学者的著作中都得到很好的阐明，但约翰·威廉·德雷珀(John William Draper)、安德鲁·D. 怀特(Andrew D. White)和詹姆斯·沃尔什(James Walsh)三人的著作尤为清晰地反映了这一点。

约翰·威廉·德雷珀(1811—1882)是杰出的美国化学家和生理学家。[23] 他因研究摄影术和参与T. H. 赫胥黎(T. H. Huxley)和塞缪尔·威尔伯福斯(Samuel Wilberforce)之间关于达尔文主义真理性的著名的1860年牛津论战而闻名于世。德雷珀是奥古斯特·孔德的信徒，晚年致力于历史研究。其《欧洲理性发展史》(*A History of the Intellectual Development of Europe*, 1863)是19世纪理性史的重要典范，而其《宗教与科学冲突史》(*History of the Conflict Between Religions and Science*, 1874)则是上个世纪读者面最广的著作。到1910年该书已出到了第25版，并且近至1972年还在重印。可能其他任何科学史著作都不能达到这个纪录。

在这两部著作中，德雷珀坚持认为科学的正当性远在宗教的正当性之上，但其第一本书中的温和腔调却在《宗教与科学之间的冲突史》中听不到了。原因何在呢？……1869—1870年梵蒂冈会议宣称教皇和教规至高无上，德雷珀对此作出如下反应：

> 让他受到诅咒吧——
>
> 那个要宣称应当用即使是在反对天启教义之时也允许人们持有他们的作为真理的主张的这样一种自由

> 精神去追求人类科学的人。
>
> 那个要宣称科学进步中除了在教会一直承认并且还在承认的意义以外的另一种意义上由教会提出的教义会实现的人。[24]

在德雷珀看来，教皇统治所渴望的就是威胁要恢复黑暗的中世纪。他认为，无异议地信仰超理性的东西的需要，结束了古代的科学进步。一个科学家需要的是某种远远不同的东西——对宇宙被永恒的规律所支配的信仰。在对真理的探求中，科学家必须向一切可能的事物敞开心扉，而不应被盲目的的信仰束缚。

安德鲁·迪克森·怀特(1832—1918)是杰出的美国外交家和教育家，他曾是美国派往德国和俄国的公使，并曾是海牙和平会议美国代表团团长(1899年)。对于我们的叙述来说，更重要的，是怀特和埃兹拉·康奈尔(Ezra Cornell)一道组建了康奈尔大学，并出任第一任校长这一事实。与当时的大部分大学相比，怀特追求的是无宗派体制，以作为各门科学和人文学科的庇护所。他既震惊又诧异地发现，这一目标遭到组织起来的各种宗教徒们的强烈反对。怀特起初希望以理性使其敌手信服，但最后，他发表了一篇题为“科学的战场”的演讲(1875年)，提出了如下论点：

> 在整个近代史中，为了宗教的想象利益而对科学加以干预，无论多么小心谨慎也许都已经产生了，而且已经并且总是导致宗教和科学二者的最不幸的灾难性后果；另一方面，一切不受限制的科学探索，无论它在

> 某些阶段看起来似乎暂时对宗教具有何等危险性，对科学和宗教同样都一直产生着无比的好处。[25]

人们对这次演讲的反应迅速而鼓舞人心。许多大学社团和文学俱乐部立即邀请他就这一主题发表演讲。这篇演讲后经扩充，写成一本小书《科学之战》(*The Warfare of Science*)出版了，并且他继续从事此项工作，就这一主题又写了“科学之战新篇章”交给《大众科学月刊》(*The Popular Science Monthly*)。在此期间，他已见过德雷珀的《科学与宗教之间的冲突》。直到他明白了德雷珀“把斗争看成是科学和宗教之间的一件事。我当时相信，而且现在确信，存在于科学和教条神学之间的就是斗争”[26]这一论述时，才第一次想到，他原来的论点必须充实。

在许多方面，怀特的历史(最终形式出版于1895年)与德雷珀的相似。他哀叹《创世纪》对科学史的影响，他论证说相信世界末日即将到来于科学的成长毫无裨益，并猛烈抨击了那些教条式的阅读《圣经》的做法。德雷珀把其工作的矛头主要对准罗马天主教会。怀特也赞成这样，但在康奈尔的经历告诉他，新教教会在这些问题上不再开明。他首先反复谈到的是启蒙运动，把它看成是理性和神秘相冲突的时期。而且，尽管科学已经努力取得了进步，但他仍认为理性尚未彻底战胜神秘。

詹姆斯·约瑟夫·沃尔什(1865—1942)的言论则对罗马天主教教义有利。他曾任福德姆医学院教授，后又任福德姆社会学院院长。在其《最伟大的13世纪》(*The Thir-*

teenth, *Greatest of Centuries*, 1907；第14版，1952）和《教皇与科学》（*The Popes and Science*, 1908）中，他把矛头直接对准怀特和德雷珀。沃尔什认为，教皇是中世纪科学和教育的重要庇护人。作为一个医生，沃尔什强调中世纪后期意大利的各个大学里进行的重要的解剖工作，并且把这种工作描绘成维萨留斯和哈维工作的必要背景。他还指出了化学和物理学中的独特工作，同时把实验方法归因于13世纪的学者们，主要是罗吉·培根和艾伯塔斯·马格纳斯（Albertus Magnus）的工作。正是教会创办了医院和大学，而相比之下基督教改革运动

> 把曾一度领先的、人们在4个世纪里获得的珍贵的一切扫荡干净。艺术、教育、科学、自由、民主——一切有价值的，当时都被毁灭。[27]

至于18世纪的启蒙运动，

> 事实是……造成了对学问和真正的教育的影响的巨大衰落。那时，人类文化明显下降。教育极为衰败，医院建筑极其糟糕，艺术和建筑学被忽视，人类自由被带上了如此沉重的镣铐，需要法国革命去解除人类肉体和灵魂的桎梏。[28]

沃尔什最早驳斥了安德鲁·迪克森·怀特的观点。至于德雷珀，沃尔什则只是把他的著作当作一部以无知为基础的"滑稽史"进行了批驳。[29]

德雷珀—怀特—沃尔什论战不仅仅是关于科学和宗教

的争论。重要的是它提倡对中世纪的科学和医学进行新的评价这一事实。在这一点上，沃尔什被证明是正确的。19世纪出版的各种中世纪医学著作得到了新的正确评价，而到新世纪的早期，许多新著作相继出版了——其中很多仍然有用——它们的作者是马克斯·纽伯格(Max Neuberger)、朱利叶斯·佩格尔(Julius Pagel)、卡尔·萨德霍夫(Karl Sudhoff)以及其他人。[30] 的确，医学史成熟于德国，并可能在第一次世界大战后仍保持了它的中心地位。同时，在物理科学方面，莫里兹·康托(Moritz Cantor)在撰写4卷本的《数学史演讲集》(*Voreslungen über Geschichte der Mathematik*，1880—1908)，保罗·坦纳雷(Paul Tannery)、托马斯·利特尔·希思爵士(Sir Thomas Little Heath)和约翰·路德维格·海伯格(Johan Ludvig Heiberg)则在准备关于希腊数学家和早期近代科学家的不朽著作。伟大的法国物理学家和科学哲学家比埃尔·迪昂(Pierre Duhem)的研究就是撰写10卷本的《世界体系》(*Le Système du Monde*，10卷；前5卷，1913—1917)，此书完全改变了人们对中世纪科学的看法，并引起了至今尚未解决的关于伽利略(Galileo)的独创性的争论。

乔治·萨顿：一个学术领域的确立

简言之，到20世纪早期，有少数学者已经清醒地认识到，科学史和医学史不仅有趣，而且对于从整体上理解历史亦十分必要。很显然，这是一个发酵的时期，正是在这一时期，乔治·萨顿（George Sarton，1884—1956）成为一名大学生。他是比利时人，分别于1906年和1911年在根特大学获得理学学士和理学博士学位。然而，他的大半生是在美国度过的，因为他在第一次世界大战中离开了自己的祖国。

尽管萨顿接受的是作为一个数学家的训练，但他对一切科学都有兴趣。他是一个具有奉献精神的人，相信自己最有价值的工作是在历史领域。为了这一目的，他参与组建了世界各地的科学史学会。他还为创办此领域的各种学术刊物而奔波。他是科学史协会和《爱西斯》（*Isis*）的创始人。该刊现在仍是此领域最著名的刊物，其第1期出版于1912年。他写下了大量的著作、论文和评论。其中最著名者当推《科学史导论》（*Introduction to the History of Science*），该书分为3卷5个部分，是上自荷马（Homer）下至14世纪的鸿篇巨著。该书的出版用了二十多年的时间，只是当萨顿认识到由于大量材料仍要被包括进来，继续写作将不可能时，他才最终放弃了完成此书的打算。萨顿还计划出版他在哈佛大学的演讲，该演讲包含了两年中他在此领域的全面研究——到他去世时，只完成了原计划8卷中的两卷。在短时间内列举出他的主要著作甚至是不可能的。对我们来说，也没有必要这么做。然而，由于他的巨大影响，谈谈他在此领域的研究态度，却是必要的。

乔治·萨顿经常表白他受惠于奥古斯特·孔德的著

作，无疑他认为自己是个实证主义者。在 1927 年的著作中，他把科学定义为“系统化的实证知识”。[31]

> 我们的主要目的，不是简单地去记录那些孤立的发现，而是要解释科学思想的进步、人类意识的逐步发展以及自觉理解和加强我们在宇宙演化中的作用的倾向。[32]

在《导论》中，他几乎没有谈及希腊人以前的古代科学，因为他觉得东方科学极其缺乏理论。另一方面，他觉得自己的工作首次对中世纪的科学作了真正的描述。这是值得怀疑的，因为迟至 1927 年，他都没有提及迪昂的《世界体系》，该书用一种萨顿没有采用的方式改变了我们对中世纪物理学的看法，直到他的哈佛演讲集的第 1 卷出版的时候(1952)，他才校正了这两个疏漏。

萨顿在其他问题上并没有止步。作为一个实证主义者，他所期望的是一门真正的科学史——即关于我们今天所知道的科学的历史。在人类较早期的认识中，除了能够构成对自然的看法的那部分外，其他东西不是被漠然置之，就是被贴上“伪科学”的标签。现在我们知道，炼金术和自然法术是近代科学发展的重要因素。萨顿愿意在其科学史中承认炼金术士们所描述的实际化学反应和设备，而不愿承认其他东西。

> 科学史学家不能把注意力过多地集中在对迷信和法术这类荒唐现象的研究上，因为那样不能帮助他很好地理解人类进步。法术本质上是落后和保守的；科

> 学本质上是进步的；前者走向没落，而后者则走向进步。除了指出这两种运动的不断冲突以及这种冲突对我们完全没有教益之外，我们即刻就无话可说了，因为它在各个时代几乎都没有什么变化。由于人的愚蠢不会立即走向进步，不会立即得到改变，也不会立即受到限制，因而，对其进行研究是一项没有希望的事业。不能激励人们去完成一个无限期的任务，去研究那种停滞的历史。[33]

第二点是，萨顿坚持科学史和医学史之间已经形成的界线。正如我们所知，19 世纪后期和 20 世纪早期，医学史得到了独立发展。萨德霍夫的学生亨利·西格里斯特(Henry Sigerist)于 20 世纪 20 年代后期离开莱比锡，前往巴尔的摩的约翰·霍普金斯大学接任新近建立的医学史研究所所长。他希望在美国继续德国的传统。但是对于萨顿来说，医学史学家们的主张对刚刚起步的科学史摆出的是一个吓人的架势。他相信存在一个科学等级系统。数学处于最高层，因为它对于天文学、物理学和化学这类数理科学是必不可少的。只有沿着这一系统，我们最终才会转而谈及生命科学。他解释说：

> 人们用不同的方式理解世界……一些人的思想较为抽象，他们很自然地首先想到的是统一性、上帝、整体性、无穷大以及诸如此类的概念，而另一些人的思想则较具体，他们深思熟虑的是健康与疾病、得益与损失。他们发明小玩意儿和药物；他们热衷于应用他们

> 所拥有的知识于实际问题，而对于了解其他任何东西均不怎么感兴趣。他们努力制造东西是为了工作和报酬以及医疗和教学。第一种人称为梦想家……第二种人则被认为是务实的和有用的。历史常常证明务实的人们目光短浅，证明“懒惰的”梦想家正确；它还证明梦想家常常被人误解。
>
> 科学史学家……不愿意使原理服从于应用，也不愿为了工程师、教师和医生而牺牲那些所谓的梦想家。[34]

萨顿的确过于崇拜梦想家。而且，由于他认为生物科学的地位远在数理科学之下，因此他认为医学也处于较低的层次。因为他确信医学是一种实用技艺，所以那些声称医学是其他各门科学的真正基础的医学史学家们使他苦恼忧伤。的确，他写道，“有关科学史的主要误解，是由于那些医学史学家们把医学看做是科学的中心”。[35]萨顿觉得，医学史学家们提出歪曲科学史的观点是由于其贫乏的科学知识所致。我们无须为医学史学界和科学史学界在今天甚至连偶尔的往来也罕见而大为吃惊……也不必为几十年中科学史以物理科学而不是以生物科学为中心而过分诧异。

我以粗略的形式考察了科学史和医学史400年的撰写史，以期作出几点一般概括。首先，显而易见的是，历史学家以某种目的从事写作——这种目的常常会成为他们根深蒂固的信条的宣传工具。这样，我们简短地考察了16世纪撰写的几部历史著作。对于帕拉塞尔苏斯信徒们来说，科学和医学真理紧密地与他们的宗教信仰联系在一起，与他们试图推翻教育法规的愿望联系在一起，与他们所信奉的帕拉塞尔苏斯医学的真理与创世中传给亚当的神授真理相关联的信条联系在一起。他们写历史是为了建立他们的医学和自然哲学。

机械论哲学在 17 世纪的成功，导致了一个新的不同的历史模式——在这个模式中宗教和科学进步相分离。那些把中世纪的科学看成是科学贫乏期的启蒙哲学家们，在自己的著作中把 1000 年的科学都抹掉了，而仅仅是踌躇地提及罗吉·培根以及其他几个在智力上似乎超出了当时普遍欠佳的水平之上的学者。的确，在这个时期，由于科学代表着人类的进步，而政治、战争和宗教则并非如此，因此科学史自觉地从各传统的历史学科中分离出来。实质上，这种研究科学的启蒙态度，完全支配着科学史进入了本世纪。沃尔什对德雷珀和怀特的历史著作的回答，较早捍卫了宗教以及中世纪的科学和医学。的确，直到 19 世纪的最后几十年，我们才开始看到一系列会改变我们对中世纪科学的看法的重要著作和历史分析。

尽管在第一次世界大战前就已经出版了许多科学史和医学史著作，但这个领域还没有在学术上得以确立，由于这个原因，乔治·萨顿才成为一个如此重要的人物。这确实应归功于他而不是其他人的努力。他不仅在哈佛大学确立了这个领域，而且还创办了《爱西斯》杂志，并且是组建国际科学史学会最认真尽责的人。他求贤若渴，鼓励那些对该领域有相同兴趣的人们继续从事研究。由于迄本世纪中期为止萨顿几十年遍及世界的活动,此领域反映出他的偏见也就不怎么奇怪了。第二次世界大战后期,少数哲学博士从哈佛大学和其他大学的这个新领域毕业了,但这些年轻学者的著作以及《爱西斯》上的文章仍然反映出强调物理科学发展的实证主义倾向。这种情况以后要根本改变。

注

[1] Morris R. Cohn and I. E. Drabkin, eds., *A Source Kook in Greek Science* (Cambridge: Harvard U. P., 1958), pp. 33-38 (Proclus), 468-73 (Celsus).

[2] 居伊·德·肖利亚克的外科学史可以在下列书籍中方便地查到:James Bruce Ross and Mary Martin McLaughlin, eds., *The Portable Medieval Reader* (New York: The Viking Press, 1973), pp. 640-49。

[3] 我详尽阐述帕拉塞尔苏斯传统的观点,见 *The Chemical Philosophy: Paracelsian Science and Medicine in the Sixteenth and Seventeenth Centuries*, 2 vols, (New York: Science History Publication, 1977)。

[4] R. Bostocke, Esq., *The difference between the auncient Physick... and the latter Phisicke* (London: Robert Walley, 1585), sig., Fiiv.

[5] Thomas Tymme, *A Dialogue Philosophicall* [London: T. S.

(nodham) for C. Knight, 1612], sig., A3.

[6] 博斯托克的历史由艾伦·G. 狄博斯附上介绍和注释重刊于"An Elizabethan History of Medical Chemistry", *Annals of Science* 18 (1962，出版于 1964 年)，1-29。

[7] 见 Debus, *The Chemical Philosophy* 1, pp. 131-34。

[8] 同上，pp. 135-39。

[9] 同上，pp. 139-45。

[10] J. Guintherius von Andernach, *De medicina veteri et noua tum cognoscenda, tum faciunda commentarij duo*, 2 vols. (Basel: Henricpetrina, 1571), 2, p. 650.

[11] 同上，pp. 26, 28, 621-22。

[12] 同上，pp. 31-32。

[13] I. 伯纳德·科恩的 *Franklin and Newton: An Inquiry into Speculative Newtonian Experimental Science and Franklin's Work in Electricity as an Example thereof* (Philadelphia: The American Philosophical Society, 1956) 提出了这样一个论点，即牛顿在 18 世纪的影响主要起源于他那得到广泛阅读的《光学》，但是几乎没有什么疑问，《数学原理》的读者群虽然不大，却为他的名声奠定了基础。

[14] Arnold Thackray, *Atoms and Power: An Essay on Newtonian Matter-Theory and the Development of Chemistry* (Cambridge: Harvard U. P., 1970), pp. 52-73.

[15] John Freind, *The History of Physick; From the Time of Galen, to the beginning of the Sixteenth Century...*, 2 vols., 4th edition (London: M. Cooper, 1750), 2, p. 204.

[16] Nina Rattner Delbart, "'Science' in Enlightenment utopias: Power and Purpose in 18th Century French 'Voyages Imaginaires,'" (U-

niversity of Chicago，博士论文，1973 年 9 月），p. 155 引用了 *the Journal de Bruxelles*（1784 年 5 月 29 日），pp. 226-27。

[17] 同上，p. 158。

[18] Joseph Priestley，*The History and Present State of Discoveries relating to Vision*，*Light and colours*（London：J. Johnson，1772）；*The History and Present State of Electricity with Original experiments*（London：J. Dodsley，J. Johnson and B. Davenport，1767）；Jean Étienne Montucla，*Histoire de Mathématiques*，*dans laquelle on rend compte de leurs progrès... jusqu'à nos jours*；*où l'on expose le tableau et le développement des principales découvertes... et les principaux traits de la vie des mathématiciens les plus célèbres*，2 vols.（Paris，1758）扩充为 4 卷（由 J. J. Le Francais de Lalande 完成和编辑，Paris：H. Agasse，1799—1802）；Jean Sylvain Bailly，*Histoire de l'Astronomie Ancienne*，*depuis son origine*，*jusquà l'éstablissement de l'école d'Alexandrie...*（Paris：Frères Dubure，1775）；*Histoire de l'Astronomie Moderne*，*depuis la fondation de l'école d'Alexandrie*，*jusqu'à l'époque de MDCCXXX*，3 vols.，（Paris：Les Frères de Bure，1779-1782）。

[19] Hermann Boerhaave，"Prolegomena，or the History of Chemistry" in *A New Method of Chemistry*；*Including the Theory and Practice of that Art*：*Laid down on Mechanical Principles*，*and accommodated to the Uses of Life. The whole making a Clear and Rational System of Chemical Philosophy*，trans. by P. Shaw and E. chambers（London：J. Osborn and T. Longman，1727），pp. 1-50. Pierre Simon de la Place，"Precis de l'Histoire de l'Astronomie" in *Exposition du Système du Monde*，3rd ed.，2 vols.，（Par-

is：Courcier，1808)，2，pp. 259-415.

[20] 关于惠威尔的研究方法，见其“序”和“导言”，第1卷，第7—11、41—51页。他坦率地声明：“这种研究之为研究，不仅仅是叙述科学史上的事实，而且是为科学哲学提供基础。”（第8页）

[21] 同上，1，pp. 185-239。

[22] 同上，2，p. 255。

[23] 唐纳德·弗莱明（Donald Fleming）的研究具有权威性。*John William Draper and the Religion of Science*（Philadelphia：University of Pennsylvania Press，1950；重印本 New York：Octagon Books，1972)。

[24] John William Draper，*History of the Conflict Between Religion and Science*，25th ed.（London：Kegan Paul，Trench，Trübner & Co.，Ltd.，1910)，pp. 350-51.

[25] Andrew Dickson White，*A History of the Warfare of Science with Theology in Christendom*，2 vols.（New York：Appleton，1900)，1，p. VIII.

[26] 同上，p. IX。

[27] James J. Walsh，*The Popes and Science：The History of the Papal Relations to Science During the Middle Ages and Down to our Own Time*（Notre Dame Edition，New York：Fordham University Press，1915)，p. 334.

[28] 同上，p. III。

[29] 同上，pp. 500-19. Appendix IX，“The Danger of Little Knowledge”（513)。

[30] 对19世纪后期和20世纪早期此领域的先驱们（莫里兹·康托、保罗·坦纳雷、卡尔·萨德霍夫、约翰·路德维格·海伯格、比爱尔·迪昂、托马斯·利特尔·希思爵士）的极有趣的概述，可以在

乔治·萨顿的"Acta atque Agenda"(1951)中方便地找到。此文重刊于 *Sarton on the History of Science*: *Essays by George Sarton*, Dorothy Stimson 选编(Cambridge: Harvard U. P., 1962), pp. 23-49。

[31] George Sarton, *Introduction to the History of Science*, 3 vols., in 5 (Baltimore: Published for the Carnegie Institution of Washington by Williams and Wilkins, 1927-1947), 1, p. 3.

[32] 同上, p. 6。

[33] 同上, p. 19。

[34] George Sarton, *A History of Science*: *Ancient Science Through the Golden Age of Greece* (Cambridge: Harvard University Press, 1952), p. XII.

[35] 同上, p. XI。

Chapter 2

第二讲

科学史：职业化与多元化

把科学作为一个整体给以更广泛的理解，并且考虑科学与人类努力的其他领域之间的相互关系，这样的综合必定大有余地。我认为，我们应该学会运用佩格尔的方法，根据发现者的全部工作去理解各个发现，然后再根据塑造该发现者的整个智识环境去理解发现者。只有这样做，我们的历史学家才可能真正地反映科学对文明的影响。只有这样做，我们才会明白科学史中传统的内在主义与外在主义这两个流派之间并不存在真正的冲突。只有这样做，科学史的价值才会得到历史学家和科学家同样的普遍接受。

在第一讲中，我讲到科学史作为一个学术性学科确立的背景——最后谈到了乔治·萨顿的几乎不朽的成就，指出了他在此领域的实证方法与更早的历史学家们的工作有很多共同之处。然而，与其同时代人及前人形成对照的是，萨顿在此领域为学者们创办了各种刊物和一个国际性学会。由于他的工作，哈佛大学形成了一项包括大学生和研究生两种水平在内的科学史项目。20 世纪 40 年代这个项目在美国授予了第一个科学史方面的哲学博士学位。

我并没有使这些演讲实际上成为我的自传这样一种意图，只因为我现在要讲的这一时期正是我开始对科学史产生兴趣的时期，所以，谈谈自己的经历，多少有点儿算是出于不得已吧。1949 年，在一位意识到科学史之重要性的历史学教授约翰·J. 默里(John J. Murray)的指导下，我在印第安纳大学撰写一篇关于 17 世纪化学的硕士论文。从事这项研究时我首次发觉了《爱西斯》和乔治·萨顿的著作。于是，在完成了化学专业的研究生学业、从事了 5 年药学工业方面的化学研究之后，我和妻子决定再度攻读研究生学位。1956 年美国在这一领域仅有 3 个研究生项目：I. 伯纳德·科恩(I. Bernard Cohen)指导的哈佛大学项目，亨利·格拉克(Henry Guerlac)指导的康奈尔大学项目以及马歇尔·克拉杰特(Marshall Clagett)指导的威斯康星大学项目。前两位是萨顿的学生;后一位是伦恩·桑代克(Lyna Thorndike)培养的,伦恩·桑代克是哥伦比亚大学伟大的中世纪史学家,因著有8卷《法术与实验科学的历史》(*History of Magic and Experimental Science*)而闻名于世。

我们决定去哈佛,我于1956年秋注册入学。仅在这几个月之前乔治·萨顿故去了。他逝世的消息已经寄到了《爱西斯》的订户手中,而当时在威德纳图书馆(Widener Libray)计划办公室里该刊仍被用作草稿纸。不久以后,他的许多著作以每本50美分之低的图书馆处理价格出售。

在这个有点神秘的领域里,学生寥寥无几,而且我们立刻发现人们当时对该学科的诠释与我们料想的有所不同。该领域的大多数新近作品是批判萨顿的,通常认为一个典型的作者就是俄罗斯科学哲学家亚历山大·柯瓦雷(Alexandre Koyré),他晚年大部分时间是在巴黎渡过的。柯瓦雷竟然坚决主张科学思想与哲学思想之间有着密切的联系,这是可以理解的,不过在他看来历史也很重要,因为只有通过历史我们才能获得一种科学观念进化的"辉煌进步"感。[1]像该领域内大多数其他学者一样,柯瓦雷把研究集中在从哥白尼到牛顿这一时期物理学和天文学的发展上。伽利略是他特别关心的一位学者,但柯瓦雷拒斥"迪昂论点"——也就是说,他不同意伽利略力学之源应该在其中世纪前辈中寻找的论点。[2]对于柯瓦雷来说,伽利略是一位远远脱离了中世纪的亚里士多德评论家们的创新者;如果实在要说他有什么前辈的话，那就只能是阿基米得（Archimedes）。柯瓦雷把科学革命解释为从亚里士多德到哥白尼的世界观的根本变革：

> 在我的《伽利略研究》（*Galilean Studies*）中，我力图定义新旧世界观的结构模式，力图确定17世纪的革命所引起的各种变化。在我看来，它们似乎应当简化成两个基本的、有密切联系的活动，我把它们描述为宇宙的毁灭和空间的几何化。[3]

这次革命不应当用社会变化来解释，它是一个从沉思冥想到积极研究的转移，或者甚至是——他补充说——“用机械模式和因果模式将目的论和有机论的思考和解释模式取而代之”。[4]在柯瓦雷看来，从多种意义上说，科学革命都是文艺复兴时期柏拉图对亚里士多德的胜利。然而，假如说萨顿不同意柯瓦雷关于柏拉图对近代科学的兴起具有重要性的观点，那么他们二人都会同意科学史学科是科学，而且认为科学史就是进步的故事。[5]

作为学生，许多百科全书式的著作被介绍给我们，这些著作以多种方式刻画了科学史的这个“英雄时代”，这些著作有伦恩·桑代克在1923—1958年间出版的8卷《法术与实验科学的历史》[6]、比埃尔·迪昂在1913—1959年间出版的10卷本《世界体系》[7]、乔治·萨顿的许多著作以及亨利·西格里斯特计划撰写的多卷本医学史中的前两卷（现已证明是仅有的两卷）[8]。此时这一领域的广阔范围似乎正在打开我们的视野。多位学者撰写、牛津大学出版的5

卷本《技术史》(*A History of Technology*),使我们深入地了解到一个被大多数科学史学家所忽视的课题。[9] 1961年李约瑟(Joseph Needham)的《中国的科学与文明》(*Science and Civilisation in China*)[10]的第1卷出版了,这是一部仍在撰写的著作,必定列为本世纪最伟大的成就之列。同年詹姆斯·里迪克·帕廷顿(James Riddick Partington)的《化学史》(*A History of Chemistry*)[11]的第1卷(实际上是第2卷)出版了。这些著作代表了学者们的毕生研究成果,这些学者相信长时期的研究能够覆盖各个领域。这是一个年轻的领域,还没有达到专业化专题著作的时代。

然而,50年代我们的学问存在裂隙这也是很明显的。在那些对伊斯兰科学感兴趣的人看来,好像不存在多少转变。我对伊比利亚半岛的科学有过特殊的兴趣,但是与伟大的航行发现的描述不同,没有什么要发现的东西。一批有奉献精神的年轻学者聚集在奥托·纽格鲍尔(Otto Neugebauer)周围,这批年轻学者正在重新发现古代近东的数学和天文学。然而,这批人相信专业化,没有做什么努力把他们的研究并入科学史的主流之中去。[12]

首先,19世纪似乎是一片荒原。I. 伯纳德·科恩于1954年所写的文章注意到:

> 一旦我们越过18和19世纪之间的界线,我们就遇不到以适合于思想史学家的某种方式写成的一般述评。梅尔茨(Merz)的早期著作呆滞、近视,而且写得拙劣,但仍然是介绍19世纪科学的主要著作。恩斯特·

卡西尔（Ernst Cassirer）的《知识问题》（*Erkenntnisproblem*）的第4卷肤浅而且对一般历史学家来说过于技术性。物理学、化学、生物学等学科的标准历史包含了大量的资料，但这些资料需要在科学思想的主流中得到整理和诠释。只有在将来才知道19世纪的科学史能否以一种在一般历史学家看来富有意义的方式加以介绍。[13]

3年后马歇尔·克拉杰特在威斯康星大学组织了一个国际性的学者小组讨论科学史的现存问题。5年后发表了研究结果，没有比这更好的著作能够表明这一领域在四分之一世纪前的状况。其中所载文章严重地倾向于各门物理科学并主要集中在中世纪后期到18世纪这一时期。克拉杰特在该论文集的序中评论说：

初看起来好像我们几乎没有强调上个世纪的发展。委员会无疑都同意这一点。但我要强调，对最后几十年的科学史进行严肃的、专门的历史研究的历史学家寥寥无几，以至于很难对这类问题的批判性讨论加以介绍。好像我们为了物理科学的发展而轻视生物学的发展。这并非我们的本意。我们最初力图集合一批著名的学者讨论19世纪的生物学，这种做法取得了部分成功。生物学史中那些可以主动研究的领域如此狭窄，以至于当我们有了一些优先取舍权时，结果我们就取消了我们原希望专门用于研究生物学所增加的工作日。[14]

事实上，自麦迪逊(Madison)会议后的几年之中，对19世纪科学的历史研究远远超过了对科学革命时期的研究。然而，这一研究不太平衡。在以进化论思想史为中心的生物学研究方面，已经迈出了很大的步伐，而物理科学史的研究很少得到综合。

与对19世纪科学研究同样重要的是，人们已经认识到科学的发展也许受到我们根本没有认为是科学的那些因素的影响。所出现的主要问题之一与艾萨克·牛顿有关。人们常常把牛顿当作一切时代中最伟大的科学家而加以颂扬，为他写传记的作者们常常有意忽视这样一个事实，即牛顿的大量文章是研究炼金术以及那些表面上似乎与经典物理学的基础和哥白尼理论的建立没有什么关系的其他问题的。更为惊人的是，人们一直忽视了帕拉塞尔苏斯和范·赫尔蒙特(van Helmont, 1579—1644)及其信徒。他们的著作在16和17世纪引起了激烈的争论，但是却被17世纪后期新的科学统治集团当作神秘的(因此也就是非科学的)东西加以拒斥。由于科学史学家们的实证主义偏见，牛顿的炼金术以及帕拉塞尔苏斯和范·赫尔蒙特的神秘主义都不是“科学”。既然科学革命的机械论哲学家正当地拒绝考虑它们，我们就应当继续这么做。

乔治·萨顿把炼金术、占星术及自然法术当作“伪科学”而未予考虑，不过如果科学史学家们选择了某种研究这一领域的不同途径，那么他这么做是成问题的。的确，历史学家——尤其是英国的历史学家，已经选择了另一条途径。

1931年赫伯特·巴特菲尔德(Herbert Butterfield)有影响的论著《历史的辉格诠释》(*The Whig Interpretation of History*)出版了，在这一论著中他论证说，实际上历史学家们已经选择了派别。他们用现代的观点编织历史，他们明显地偏爱16和17世纪的新教改革者，他们用这种观点来定义“进步”。从政治上看，他们撰写“辉格的”历史是有愧的。这些历史学家们觉得需要作出结论，而这样一来他们就使原始材料的丰富多彩的复杂性过于简单化。他写道：

> 历史的价值在于恢复其过去丰富多彩的活生生的生命。它是不能用干巴巴的线索叙述，也不能用某种几何形式阐明其意义的故事。好像在井底有某种绝对的东西，有某种独立于时间或环境的真理，通过蒸发人类和个人的因素，偶发的或瞬间的或局部的事件以及偶然的成分，能够得到某种历史的精髓，然而这种精髓是不存在的……[15]首先，历史学家的作用不是得出该称之为价值判断的东西……[16]他的作用是描述，他公正地站在基督教徒和伊斯兰教徒之间，他对任何一种宗教都不感兴趣，除非宗教与人类生活有牵连……[17]当他使我们脱离简单而绝对的判断并使一切都恢复到相互牵连的历史的来龙去脉之中时，他就回到了他本来的位置。当他告诉我们根据环境，根据所产生的相互影响来判断一件事是好是坏时，他就回到了他本来的位置。如果历史能做到这一切，它就会使我们想起那些削弱了必然事件之基础的错综复杂的情况，并向

> 我们表明，我们的所有判断都只与时间和环境有关。[18]

巴特菲尔德的“宣言”是对所有历史学家的挑战。事实上，他只是后来才对科学史特别感兴趣，我们待会儿还要谈到他。

在科学史学家和医学史学家当中，首先使人们注意到那些被忽视的历史人物的是沃尔特·佩格尔(Walter Pagel)。[19]但是，尽管他关于范·赫尔蒙特的第一部著作1930年就出版了，而其广为流传的方法论影响，还只是后来注明日期出版了他的《帕拉塞尔苏斯》(*Paracelsus*, 1958)以及《威廉·哈维的生物学思想》(*William Harvey's Biological Ideas*, 1967)之后的事。佩格尔认识到萨顿“逐步揭示真理的历史”的谬误，并辩驳说，“这样一种以现代观点选择材料为根据的方法可能危及历史真相的描述”。[20]的确，“站在现代科学和医学的立场上，从原初的来龙去脉中取出过去的发现和理论加以评价的各种历史”，很有可能危险地误入歧途。[21]

那么，科学史学家该如何研究呢？佩格尔谈到自己的研究时曾提出：

> 历史学家应该力图弄清过去在其他方面“健全的”科学工作者们哲学的、神秘的及宗教的“侧面台阶”，而不应该选择那些在近代科学的侍僧看来“讲得通”的材料。正是这些“侧面台阶”向历史学家提出了挑战：要揭示它们在学者心目中存在的内在理由和原委以及它

们与该学者的科学观念之间的有机联系。换言之，就是要颠倒历史学家的科学选择方法并在原来的境况中重新陈述他的英雄思想。两套思想——科学的和非科学的——并不是简单并列地或彼此无关地出现的，而是一个相互支持、相互确证的有机整体。没有任何其他方式向我们昭示学者的面貌。[22]

把医学史和科学史的事实诠释为“发生这些事实的时代的外部表现”一直是佩格尔的愿望。他解释说，他的这一愿望实现之时，

似乎就可以看出，不仅技术装置的可靠标准使发现成为可能，而且这些标准也可看做是完全非科学的观念和某种特殊文化背景的产物。……医学史似乎比起医学用通常所使用的对进步的直线式透视所显示的结构要复杂得多。然而，如果我们想写历史——并且是极不畅销的历史，那么我们就不得不着手重建古代思想这一繁重的工作。[23]

佩格尔曾经告诉我，他在听了柯瓦雷关于牛顿物理学的演讲后站起来询问牛顿的炼金术著作。柯瓦雷对此不予考虑，他说，“我们不涉及这个问题。”从他的观点看，他是对的，但对于佩格尔来说，除非我们考察牛顿的全部著作，否则就不可能理解这个“完整的人”。或许没有比这件轶事能更好地说明这两位学者之间基本差异的了。

佩格尔的著作虽然重要，但他的影响也许没有其后的戴姆·弗朗西斯·耶茨(Dame Frances Yates)的影响大，耶

茨撰写了一系列有关科学革命与炼金术关系的著作。戴姆·弗朗西斯是一位文学史学家,她于1964年出版的《乔达诺·布鲁诺与炼金术传统》(*Giordano Bruno and the Hermetic Tradition*)一书引起了科学史学家们的注意。[24] 书中试图把布鲁诺作为一位16世纪日心说的支持者来评价其著作,这不是因为他是一位有远见的科学家,而是因为以太阳为中心的系统极好地容纳了他关于太阳和宇宙的神秘的、"炼金术的"观点。该书的确是过去20年中出版的科学史方面最有影响的著作之一。总的来看,这一影响是健康的,因为她极力主张史学家们认真对待过去绝不应该忽视的大量著作。

耶茨的影响也有其危险的一面。由于过分强调炼金术、新柏拉图主义、法术以及文艺复兴时期哲学的其他神秘倾向的重要性,她提出了更多的越来越没有确凿证据的大胆主张。在《罗齐克鲁会启蒙运动》(*Rosicrucian Enlightenment*,1972)一书中,她坚持认为整个科学革命是由文艺复兴时期的神秘主义和法术发展而来的。[25] 她力求把伦敦皇家学会的发源以及笛卡儿(Descartes)和牛顿的工作与约翰·狄伊(John Dee)及该世纪早期的玫瑰十字会文献联系起来。[26] 遗憾的是,这些联想没有得到所需的可靠的历史证据的支持。它们充其量不过是一些值得怀疑的推测。

佩格尔和耶茨的著作引起了极大的兴趣和争论。也许值得注意的是,他们都没有对科学史中老的、传统的领域作过任何阐述。最好是把耶茨当作文学史学家,而佩格尔则是一位古典主义者、病理学家和医学史学家。但是，他们都向科学史提供了一个挑战性的方法——也许大有助于解决科学革命的整体问题的方法。

在科学史学家当中，对于伪科学的研究激起了与艾萨克·牛顿著作的恰当诠释有关的激烈论战。所有的人都会同意，牛顿代表了近代早期物理学、数学及天文学中许多思想的顶峰，但如何诠释他所写下的数千页关于炼金术的手稿呢？首先试图把它们综合为一幅完整的牛顿画像的是R. S. 韦斯特福尔（R. S. Westfall），他早期根据传统的科学内史讨论了这位学者。到了20世纪70年代早期，韦斯特福尔便确信17世纪赫耳默斯神秘主义是牛顿思想的一个基本成分，而且它“能使相对粗糙的17世纪科学的机械论哲学达到更高的完善程度”。

> 牛顿思想中的炼金术成分与科学事业终归不是对立的。恰恰相反，通过赫耳默斯传统和机械论传统二者彼此的结合，他建立了一个家系，这个家系自称那种今天不理解地鄙视与炼金术有关的各种玄秘思想的真正科学就是其直系后裔。[27]

最近在牛顿炼金术的研究中作出贡献的B. J. T. 多布斯（B. J. T. Dobbs）进一步主张牛顿的许多极重要的著作源自他的炼金术思考，而且“从某种意义上说，他1675年以后的全部经历都可以看成是试图使炼金术和机械论哲学合二为一的生涯”。[28]

更传统的科学史学家们对这些新发展表示忧虑，这并不奇怪。在剑桥大学国王学院的一次专门研究该领域新动向的会议上(1968年),P. M. 拉坦西(P. M. Rattansi)论证说,历史来龙去脉的真相表明,“历史学家们的任务不能是把‘合理的’和‘不合理的’成分分离开来,而是把它作为一个整体并在具有相当深度的探索基础上寻找冲突之点和紧张之点”。[29]在答辩中,马丽·赫斯(Mary Hesse)反对把那些按照现代看法不算是真正科学的主题纳入这个领域。伪科学也许属于历史领域,但不能认为它们是科学史的组成部分。希望把这样的做法看成是唯一的做法,她补充说,十分必要的是,我们应该把现代科学作为衡量过去论点的工具。用掺杂着非科学因素的过去进行评价只能浪费我们的时间。她断定,由于“把更多的光投射在画面上会歪曲已经看见的东西”,[30]我们的确必须小心对待我们所阅读的或允许我们自己评价的东西。赫斯的反应是更传统的科学史学家和科学哲学家当中最极端的。

赫斯和拉坦西之间的交锋明显地陷入僵局,这表明了该领域目前的紧张状态。然而,所谓伪科学的作用未必就是这种状态的唯一根源。或许此刻最激烈的争论涉及的是科学与社会的关系。仅在几年前,这似乎还不是一个怎么重要的问题。托马斯·S. 库恩(Thomas S. Kuhn)为《社会科学百科全书》(*Encyclopedia of the Social Science*, 1968)撰写了科学史条目,他将“内在主义的”和“外在主义的”科学史加以比较。前者研究与科学发展有关的技术问题;后者则“试图将科学置于一种文化的史境中,以加深对科学发展及其影响的理解”。[31]特别有趣的是关于罗伯特·金·默顿(Robert King Merton,1938)论点的争论,默顿试图通过指出以下两点来解释17世纪英国科学的成功:(a)培根强调的实用技艺和商业过程;(b)宗教方面清教主义的刺激。[32]如果库恩主张科学内史和科学外史是互补的,他也会觉得在很大程度上这是那些没有成功地证明自己观点的学者们提出的一个旧论点。至于“新一代历史学家”——即大多是柯瓦雷唤起的历史学家们——他们

> 声称已经表明16和17世纪对天文学、数学、力学乃至光学的基本的修正,不应该归功于新的仪器、实验或者观察。他们论证说,伽利略的主要方法是使烦琐科学趋于完善的传统的思想实验。[33]

这已经远远脱离了技艺传统和培根那一贯失败的新方法论。就所涉及的17世纪而言,他认为,只有电学、磁学、化学及热现象研究之类的“新”科学借用了技艺传统。[34]各门数理科学则应当继续用种种内在方法进行研究。

库恩那备受称赞的《科学革命的结构》(*Structure of Scientific Revolutions*,1962)是试图用一个科学范式代替另一个科学范式来解释科学革命的内在主义的研究。[35]由于人们对科学成长中非科学因素的兴趣不断增长,这部著作并不像人们所预料的那样对科学史学家们产生很强的影响。相反,它受到的是社会科学家、哲学家和历史学家们的欢迎,这些人不是把它当作科学史模型来使用,而是用它来考察他们自己领域的内部发展。[36]

在托马斯·库恩看来,“新的”科学史主要是内在主义的。而60年代后期和70年代早期人们越来越对科学与社会的关系感兴趣。由于这个原因,科学史成了对历史学家、哲学家和社会学家很有吸引力的一个领域——他们中的许多人在科学或科学史方面没有受到过什么训练。这些作者争辩说,没有那些以前看来似乎是必需的专门科学知识,现在也可以掌握科学史的各个重要方面。从那时起,由此一直搞混了结果,事实上,专业知识甚至在科学史的一些最神秘的领域仍然很重要。大量重要的研究仍然一直在发表。例如,基思·托马斯(Keith Thomas)的《宗教与法术的衰落》(*Religion and the Decline of Magic*,1971)对我们了解英国近代早期的知识舞台就是一个不朽的贡献。[37]克里斯托弗·希尔(Christopher Hill)的著作也很重要,在他的《颠倒的世界》(*The World Turned Upside Down*,1972)一书中,他把对炼金术和帕拉塞尔苏斯信徒的最新研究作为他了解英国内战①的关键。[38]

在《1689—1720年的牛顿信徒与英国革命》(*The Newtonians and the English Revolution*,1689-1720,1976)一书中,玛格丽特·雅各布(Margaret Jacob)论证说,牛顿物理学的胜利也许不应当归因于牛顿科学的价值,而应当归因于这样一个事实,即“光荣革命”(1688年)时期,英国神学家通过对牛顿的综合的支持寻找一个有力的同盟者。

---

① 指1642—1649年间查理一世与议会的战争。

> 这些教士们……用新的机械论哲学支持基督教，攻击无神论，从而传播新科学的思想及与之相随的自然哲学。没有最初的不拘泥于教条及形式的人们喋喋不休的说教，科学仍然只能秘传甚至还可能使那些受过教育而又虔诚的大众感到恐惧。[39]

她把新科学看作是对霍布斯和笛卡儿的旧机械论哲学以及各大学的亚里士多德哲学和中世纪的各种激进宇宙论的明确的抵制，而新科学思想也常常与造教会和国会反的那些观点联系起来了。

> 当然，明显的问题是为什么这些开明的教士觉得非要抵制一种自然哲学的科学而接受另一种不可。科学史学家们常常相信，机械论哲学在英国取得成功完全是因为它提供了对自然的最花言巧语的解释。也许正是那样，但在我对使之能够接受的历史进程的理解中，新机械论哲学与自然秩序的实际行为之间想象的一致并不是新机械论哲学早期成功的主要原因。[40]

对于雅各布来说，牛顿主义的成功可以从社会方面解释为“英格兰圣公会的智力领袖们把它作为他们喜欢称为‘世界政治活动’的那个东西的幻想基础来使用。牛顿那由命定的、靠天佑控制且受数学支配的宇宙为被自私自利的人们所控制的稳定昌盛的政体提供了一个模型。”[41]简言之，我们在这里看到了以完全脱离事实为基础对牛顿学说的成功作出的一种解释，这个事实就是，牛顿的工作代表了从哥白尼的《天体运行论》(*De revolutionibus robium*，1543)到《数学原理》(1687)将近一个半世纪科学讨论和科学研究所达到的顶峰。

美国科学促进会1979年12月举行的一次会议上,查尔斯·C. 吉利斯皮(Charles C. Gillispie)痛斥了该领域里追随较新潮流的人。正如《科学》报道的,吉利斯皮抱怨说:"科学史正失去其对于科学的把握,正严重地依赖于社会史,正遭到冒牌学问的戏弄。"他抨击了那些讨论科学问题但却很少或根本没有受过科学训练的人。

> 对吉利斯皮来说,不怎么可憎但却讨厌的是完全忽视科学的社会史,例如涉及妇女在某个特殊科学协会中的作用但却不管她们的实际科学工作的那类研究……他说,另一种倾向是学者们把焦点集中在个人或轶事上:研究牛顿的炼金术而不是他的运动学,研究凯库勒(Kekule)的蛇舞而不是他的苯环,研究达尔文(Darwin)的神经疾病而不是他对证据的整理。有些所谓的学者只注意丑闻……吉利斯皮说:"这些学者对我们现在所从事的科学中最严格地排除了的那类事情——那类荒谬的、个人的事情——怀有一种不健康的欲望。"[42]

吉利斯皮为回到柯瓦雷的价值准则上去所作的抗辩已被社会史学家们驳回,这些社会史学家们说:

> 迄今为止,科学社会史作为一种处理过去的合法的方法,在学科之内已经自我确立了。尽管有C. C. 吉利斯皮最近的这场后卫战,但大多数历史学家都承

认，分析科学内部理论发展的传统做法，需要用对科学活动的社会基础变化的研究来加以补充。20 世纪 60 年代晚期的各场“内外”之争是人们对过去所寄托的希望。[43]

目前可以看到，在医学史上存在着类似的一场争论，过去人们常常认为此领域的学者在取得医学史博士学位之前必须获得医学博士学位。他的研究应该集中在医学理论和疾病的科学方面。这种传统仍在继续，但不会再有优势。今天医学史之利刃所对准的是医学的社会背景，这些历史学家们通常是通过设在历史系而不是设在老的医学史系或医学院里的项目得到训练的。由于这些学者们没有受到其前辈的医学的训练，我们在各种学术刊物上现在仍可见到与这一主题以及适当的医学史训练有关的争论。[44]

乔治·萨顿逝世时科学史是作为一个小领域建立的，但人们已认识到它的重要性。然而，由于它的历史发展，科学史通常以独立于历史或科学的项目形式在学术界出现。大多数在 25 年前发表著作的科学史学家已受到了科学家的训练。萨顿认识到了这一点，但却认为，将来职业科学史学家在获得科学史方面的哲学博士学位之前，至少应该取得两个硕士学位——一个科学硕士学位和一个历史学硕士学位。然而，柯瓦雷的影响以及哲学家中脱离哲学史转向科学哲学的潮流使科学史与科学哲学中的独立项目的发展得到加强。在五六十年代，关于科学史与历史和科学之间的关系有过更进一步的讨论。

1956年，科学史需要某种专门的科学知识似乎已很明显，这种训练似乎与所有的但不是最非凡的历史学家所接受的训练不一样。但此时传统的历史学家们已开始意识到科学与技术对于我们生活的惊人影响，这样，从该领域学习更多的东西也就显得迫切了。因此，1959年赫伯特·巴特菲尔德在关于"科学史与历史研究"的一次讲演中说道：

> 尽管世人始终知道科学和技术重要，但只是在最近，这些东西才开始支配我们的命运——那种我们以前从历史书本中学到的认为是极大地依赖于政治家们的意志的命运。[45]

他论证说，历史学家们必须重视近代科学的兴起，当他们这样做的时候，就会"改变史学的整体性质"。[46]但是，巴特菲尔德对科学史的独立性并未表示异议。在他的很有影响的《近代科学的起源，1300—1800》(*The Origins of Modern Science, 1300-1800*，1949)一书中，他提出了就在战后的那几年里流行的研究该领域的常规实证方法。[47]科学史必须通过历史学家来理解，但由于它需要专业化知识，它能够恰当地自行发展。事实上，历史学家们注意到了巴特菲尔德提出的更多地认识科学的号召。当20世纪六七十年代科学史领域中授予了越来越多的博士头衔时，这些年轻学者中的绝大多数发现他们本人应该受雇于历史系而不是受雇于老的独立的科学史项目或者科学史与科学哲学项目。传统历史学家们的这种新兴趣确实加快了奔向新研究领域的步伐，这些新领域有如我已经指出的那样，包括伪科学在近代科学兴起中以及与科学、社会和文化之间的关系有关的更一般的问题中所起的作用。在最近几十年里，该领域的发展又重新开始了科学史与各门科学之间关系问题的讨论。20 世纪 50 年代，在论证科学教育的改革方面几乎没有比詹姆斯·B. 科南特(James B. Conant)更有影响的科学家。这次论战清楚地表明，需要对美国青年进行更先进的科学训练。结果，重新考虑了各门科学的教学方法——而同时，为了给主修非自然科学专业的大学生提供一个了解各门科学如何发展的机会，引进了历史的“案例研究”。但是，科南特在发表于 1960 年的一次讲演中陈述道，历史恰好对于科学家来说具有价值。他认为科学教育通常过于狭窄，使用历史案例方法可以使学生的视野更广、获得的信息更多。[48]他拟出了一个新的科学课程表，让学生首先接受本专业的科学史训练，然后再接受近代科学史的训练。这些课程之后，紧接着开设的是在尽可能广的意义上的科学史方面的其他课程——只有这时——才能与以前的科学史课程联系起来理解文化史和政治史。由于科学史对各门科学的价值，科南特为建立实力雄厚的科学史系而四处呼吁。他对企图把科学史与科学社会史等同起来，或者与科学哲学等同起来的那些人热情甚少。[49]

虽然 50 年代和 60 年代美国大学里广泛采用了研究科学的历史案例法，但科南特对科学家进行历史训练的雄心勃勃的计划却从未结出果实。然而，这些课程终于大都停开了。对于科学家来说，这些课程要讲授大量的资料似乎进度太慢。对于科学史学家们来说，其中的历史又不够——对于那些对科学不怎么有兴趣的学生们来说，这些课程通常没有比集中地概述某门特定的科学更有趣。[50]

某个科学家掌握的大量早期文献知识导致某个突破，尽管在科学史中有几个这样的著名例子，但这种情况是极为罕见的。我想到1950年巴特菲尔德主持的一场讨论。大约有30名学生和教员在场，巴特菲尔德问是否有人知道科学史知识证明是对某个科学发现具有直接价值的事例。我是在场中唯一回答问题的——很有可能是因为我在不久前研究了惰性气体的发现，并且知道拉姆塞（Ramsay）阅读了卡文迪什（Cavendish）在18世纪所写的文章以及这些文章对于拉姆塞的影响。

托马斯·库恩同意，阅读科学史也许不会使各门科学从中获得很大益处。他在1968年写道：

> 在与科学史有关的各个领域中，可能影响最少的就是科学研究自身。科学史的倡导者们不时地把他们的研究领域描绘成一个被遗忘的思想和方法的丰富宝库，其中有一些可以很好地解决当代的科学困境。当一种新概念或新理论在一门科学中成功地展开的时候，常常在这一领域的早期文献中发现某些以前被忽略的先例。这很自然地会使人怀疑注意历史究竟是否能加速革新。然而，几乎可以肯定，答案是不能。供研究的材料数量的有限、合适的分类索引的缺乏以及在

预期与实际改革之间难以捉摸的巨大差别，所有这一切结合起来即可提示，科学新事物最有效的来源仍然是重新发明而不是重新发现。[51]

一般说来我会同意库恩的结论：……我们研究科学史不应该只是希望寻找那些今天仍然有效但早已被遗忘了的发现的规律。但是，科学史对科学家和非科学家来说确有实际价值。通过对自己领域的早期文献的研究，学生就会了解科学的进程。他确实应该懂得我们今天所接受的结果一般说来并不是以简单的方式获得的这一事实——懂得同样的争论过程同样也是今天的科学的组成部分。我曾经听说天文学史学家们论证天文学史知识确实帮助澄清了现代的天文学争论。我们发现研究达尔文进化论的历史学家们不仅在生物科学的论战中而且在激烈的上帝创世论的论战中起着积极作用。

在过去的一小时中，我有选择地勾勒了这个迅速成长的领域的变化，仅仅集中论述了以下几个因素：其题材的发展，在近代科学发展中引入所谓伪科学和社会因素而引起的论战，以及关于科学史与历史和各门科学之间关系的讨论。也许我应该转向其他论题，比如人们对于技术史的新兴趣及其在特殊的民族背景中与科学研究的关系，但我认为我所选择的例子足以表明过去几十年里专家们所研究的各种类型的问题。

有时我感到纳闷，如果萨顿仍然在世的话，他对这个领域会有什么想法。在他逝世时他的实证主义观点仍占统治地位。它们今天却过时了。科学史在一个较短的时期内发展到了现在这种状态，但这种成熟状态也已经伴随着与这种成熟相一致的惬意的进步感的衰退。的确，某些科学史学家现在所取的守势态度正好表明对各种新方法论的兴趣已经多么深地渗透到该领域。今天，除了科学家和科学史学家，还有许多历史学家、文学批评家和社会科学家正在有效地将科学史应用于他们自己的研究领域。

然而，真的无须担心——像吉利斯皮明确表示的那样——科学史将失去其对于各门科学的技术资料的需要。科学史将一直需要在各门科学中进行技术性的研究，并且因此而产生的历史总会有价值。但这并不意味着我们不愿意超越这个领域内代表战后的许多学术成就的技术性专题研究以及技术性批评。把科学作为一个整体给以更广泛的理解，并且考虑科学与人类努力的其他领域之间的相互关系，这样的综合必定大有余地。我认为，我们应该学会运用佩格尔的方法，根据发现者的全部工作去理解各个发现，然后再根据塑造该发现者的整个智力环境去理解发现者。只有这样做，我们的历史学家才可能真正地反映科学对文明的影响。只有这样做，我们才会明白科学史中传统的内在主义与外在主义这两个流派之间并不存在真正的冲突。只有这样做，科学史的价值才会得到历史学家和科学家同样的普遍接受。

注

[1] Alexandre Koyré, *Études Galiléennes*, 3 parts, 1935-1939；后合为一卷重印（Paris：Hermann，1966），p. 11。

[2] 尤见 Pierre Duhem, *Le Système du Monde*, 10 vols.，（Paris：Hermann，1913—1959）.

[3] Alexandre Koyré, *From the Closed World to the Infinite Universe*（New York：Harper Torchbook，1958），p. VI.

[4] 同上，p. V。

[5] 萨顿关于柏拉图的论述见 *A History of Science Through the Golden Age of Greece*（Cambridge：Harvard U. P.，1952），pp. 395-430. 他讨论了柏拉图与《蒂迈欧篇》有关的科学思想，他写道："《蒂迈欧篇》对后世的影响巨大，十分有害……《蒂迈欧篇》的科学堕落被错当成科学真理。也许除了圣约翰（St. John）和迪万（Divine）的《启示录》，我找不到比这部著作影响更坏的其他任何著作来。然而，人们认为《启示录》是一部宗教著作，而《蒂迈欧篇》是一部科学著作；错误和迷信只有当它们在科学的外衣下被销售给我们的时候才最危险。"

[6] Lynn Thorndike, *A History of Magic and Experimental Science*, 8 vols.（New York：Columbia University Press，1923-1958）.

[7] 见[2]。

[8] Henry Sigerist, *A History of Medicine*, 2 vols.（New York：Oxford U. P.，1955，1961）.

[9] Charles Singer, E. J. Holmyard and A. R. Hall, eds., *A History of Technology*, 5 vol. (New York and London: Oxford U. P., 1954—1958).

[10] Joseph Needham, *Science and Civilisation in China*, *Volme* 1: *Introductory Orientations* (Cambridge: Cambridge U. P., 1961).

[11] 首先出版的是论述16和17世纪的第2卷。J. R. Partington, *A History of Chemistry* 2 (London: Macmilan, 1961)。

[12] "我对于进行'综合'——无论这个术语可能有哪一种含义——的任何努力都极为怀疑,我相信专业化是健全的知识的唯一基础。" O. Neugebauer, *The Exact Sciences in Antiquity* (1952; New York: Harper Torchbooks, 1962), pp. V-VI.

[13] I. Bernard Cohn, "Some Recent Books on the History of Science", *Roots of Scientific Thought*: *A Cultural Perspective*, Philip P. Wiener and Aaron Noland, eds. (New York: Basic Books, 1957), pp. 627-56 (656). 最初发表于 *Journal of the History of Ideas*。

[14] Marshall Clagett, ed., *Critical Problems in the History of Science*: *Proceedings of the Institute for the History of Science at the University of Wisconsin*, *Septenber* 1-11, 1957 (Madison: The University of Wisconsin Press, 1962), p. VI.

[15] Herbert Butterfield, *The Whig Interpretation of History* (193; New York: W. W. Norton & Co. Inc., 1965), p. 68.

[16] 同上, p. 73。

[17] 同上, p. 74。

[18] 同上, p. 74-75。最近对巴特菲尔德及"辉格"史的讨论,见 A. Rupert Hall, "On Whiggism", *History of Science* 21 (1983), pp. 45-59。

[19] Walter Pagel, *Jo. Bapt. Van Helmont*: *Einführung in die philos-*

ophische Medizin des Barock (Berlin: Springer, 1930); *Paracelsus: An Introduction to Philosophical Medicine in the Era of the Renaissance* (Basel / New York: S. Karger, 1958); *William Harvey's Biological Ideas: Selected Aspects and Historical Background* (Basel / New York: S. Karger, 1967).

[20] Walter Pagel, "The Vindication of Rubbish", *Middlesex Hospital Journal* (Autumn, 1945), pp. 1-4 (1).

[21] 同上。

[22] Pagel, *Harvey's Biological Ideas*, p. 82.

[23] Pagel, "The Vindication of Rubbish", p. 4.

[24] Frances A. Yates, *Giordano Bruno and the Hermetic Tradition* (Chicago: The University of Chicago Press; London: Routledge & Kegan Paul; Toronto: The University of Toronto Press, 1964).

[25] Frances A. Yates, *The Rosicrucian Enlightenment* (London: and Boston: Routledge & Kegan Paul, 1972).

[26] 同上，pp. 113, 171-205。

[27] Richard S. Westfall, "Newton and the Hermetic Tradition", in Allen G. Debus, ed. *Science, Medicine and Society in the Renaissance: Essays to honor Walter Pagel*, 2 vol. (New York: Science History Publications, 1972) 2, pp. 183-198 (195).

[28] Betty Jo Teeter Dobbs, *The Foundations of Newton's Alchemy or The Hunting of the Greene Lyon* (Cambridge / London / New York / Melbourne: Cambridge U. P., 1975), p. 230.

[29] P. M. Rattansi, "Some Evaluations of Reason in Sixteenth and Seventeenth-Century Natural Philosophy", *Changing Perspectives in the History of Science: Essays in Hornour of Joseph Needham*, eds. Mikulás Teich and Robert Young (London: Heinemann,

1973), pp. 148-66 (150).

[30] Mary Hesse, "Reasons and Evaluations in History of Science", 同上,pp. 127-147 (143)。

[31] Thomas S. Kuhn, "History of Science", *International Encyclopedia of Social Sciences*, ed. David L. Sills, (New York: Crowell Collier and Macmillan, 1968, 1979), 14, pp. 75-83 (76-82). ["科学的历史",托马斯·S. 库恩著,纪树立、范岱年、罗惠生等译,《必要的张力》,第103—126页,福建人民出版社,1981年。]

[32] 同上, pp. 79-81。

[33] 同上, p. 80。

[34] 同上。

[35] Thomas S. Kuhn, *The Structure of Scientific Revolution* (Chicago: The University of Chicago Press, 1962). [T. S. 库恩著,纪树立、李宝恒译,《科学革命的结构》,上海:上海科学技术出版社,1980年。]此书亦作为芝加哥大学出版社出版的《国际统一科学百科全书》之第2卷第2号发表。

[36] 作为这种例子,可见下列文献:Barry Barnes, *T. S. Kuhn and Social Sciences* (New York: Columbia University Press, 1982); Signe Seiler, *Wissenshaftstheorie in der Ethnologie: zur Kritik u. Weiterführung d. Theorie von Thomas S. Kuhn anhand ethnograph* (Berlin: Reimer, 1980); Garry Gutting, ed., *Paradigms and Revolutions: Appraisals and Applications of Thomas Kuhn's Philosophy of Science* (Notre Dame: University of Notre Dame Press, c. 1980)。

[37] Keith Thomas, *Religion and Decline of Magic: Studies in Popular Beliefs in Sixteenth- and Seventeenth-Century England* (1971; Harmondsworth: Penguin, 1973).

[38] Christopher Hall, *The World Turned Upside Down: Radical Ideas During the English Revolution* (1972; New York: The Viking Press, 1973), 尤见 pp. 231-46。

[39] Margaret C. Jacob, *The Newtonians and the English Revolution 1689—1720* (Ithaca: Cornell U. P., 1976), pp. 16-17.

[40] 同上, p. 17。

[41] 同上, p. 18。

[42] William J. Broad, "History of Science Losing Its Science", *Science* 207 (January 25, 1980), 389.

[43] Paul Wood, "Recent Trends in the History of Science: The dehumanisation of History", *BSHS Newsletter*, No. 3 (September, 1980), 19-20 (19).

[44] 见 Leonard G. Wilson, "Medical History Without Medicine", *Journal of the History of Medicine and Allied Sciences* 35 (1980), 5-7; Lloyd G. Stevenson, "A Second Opinion", *Bulletion of the History of Medicine* 54, (1980), 135; Ronald L. Numbers, "The History of American Medicine: A Field in Ferment", *Reviews in American History* 10, (1982), 245-64; Gert H. Brieger, "The History of Medicine and the History of Science", *Isis* 72 (1981), 537-40。

[45] Herbert Butterfield, "The History of Science and the Study of History", *Harvard Library Bulletin* 13 (1959), 329-47 (329).

[46] 同上, p. 347。

[47] Herbert Butterfield, *The Origins of Modern Science 1300—1800* (New York: Macmillan, 1952). [赫伯特·巴特菲尔德著，张丽萍、郭贵春等译，《近代科学的起源》，北京：华夏出版社，1988年。]

[48] James B. Conant, "History in the Education of Scientists", *Harvard Library Bulletin* 14 (1960), 315-33 (322-23).

[49] 同上，p. 325。

[50] 这是我本人于1957—59年以及1961—63年间在哈佛大学和芝加哥大学讲授了4年这种形式的课程之后作出的估价。

[51] Kuhn, "History of Science", p. 81.

Chapter 3

第三讲

化学史的意义

化学史的撰写史至少已有400年，但仍然还有许多事情要做。确实，我们现在有很多理由可以说明重写内部史的必要性，但我们在智识史、政治史和社会史方面对化学和化学技术的作用仍然知之甚少。我认为，不理解化学所起的作用，就不能理解科学革命。而且我相信，在我们将来所知比现在所知更多之时，我们还会说，没有化学史知识，就不可能理解世界现代史。

对化学专业的学生来说，化学史的撰写史在某种程度上是一个谜。作为一个职业化学家，我本人借以学习化学的那些教科书均由历史导论开篇。我所就学的系，都开设了化学史课程，这在当时还是不常见的。[1]这类课程吸引不了多少学生，但似乎应该确立这个领域。然而当我在哈佛大学做学位论文打算研究化学史时，我又略感有些叛逆之嫌，那里的科学史学家们都置化学史于不顾。科学通史课程仅仅涉及拉瓦锡(Lavoisier)，而在研究生的物理科学史讨论会上，则完全不涉及化学。当时我非常吃惊，但只要回顾一下历史，答案就清楚了。

作为一个化学专业的学生，我正确地认为，这门学科中有很强的历史传统。今天这一习惯也许不那么普遍，但20世纪化学教科书的大多数作者们，都感到非要在他们的著作的开头做一点历史介绍不可，他们保持的这种习惯也许可以上溯数百年。我就是通过使用帕廷顿的《无机化学教科书》(*Text-Book of Inorganic Chemistry*)[2]以及斯尼德(Sneed)和梅纳德(Maynard)的《普通无机化学》(*General Inorganic Chemistry*)[3]开始了解化学史的，这两部书都有介绍历史的专章。上世纪末仍在使用的罗斯科(Roscoe)和肖莱马(Schorlemmer)的《论化学》(*Treatise on Chemistry*)，开头40页讲的就是历史[4]，而拉姆塞、贝托雷(Berthelot)、柯普(Kopp)以及肖莱马的名字则使人想起权威的历史成就和化学成就[5]。到19世纪初，用化学史来介绍有关化学的一般知识已经成为通常的习惯，以至威廉·亨利(William Henry)提出，撰写一部教科书如果没有这种规范的特征的话，就必须解释一下省略的原因。[6]约瑟夫·布莱克(Joseph Black)18世纪后期关于化学的演讲，表明他很早就接受了拉瓦锡的化学学说，这些演讲也因此而著称于世。[7]出版的演讲稿中没有科学史，但是道格拉斯·麦克凯(Douglas McKie)通过研究这些演讲的手稿，指出它们也是用历史介绍开头的。从布莱克时代到现代的许多例子，都可以用来说明化学著述中这种对化学史的持续不衰的兴趣。

提到布莱克之前的时期，人们立刻想起波尔哈夫对这门学科的发展所作的介绍性概述，尽管布莱克的演讲深受赫尔曼·康林吉乌斯(Hermann Conringius，1606—1681)与奥劳斯·博里齐乌斯(Olous Borrichius，1626—1690)之间的历史论战的影响。康林吉乌斯是当时最博学的文豪之一，他在黑尔姆施泰特(Helmstedt)任教并于1648年出版了《论古埃及人的炼金术医学与帕拉塞尔苏斯主义者的新医学》(*De hermetica Aegyptiorum vetere et Paracelsiciorum nova Medicina*)一书。作为一位亚里士多德信徒和盖伦主义者，他反对帕拉塞尔苏斯信徒们的化学论医学，并否斥赫耳墨斯·特里斯麦基斯托斯(Hermes Trismegistus)①的历史存在。博里齐乌斯是哥本哈根的一位化学和植物学教授,他于20年后在其《论化学的起源与进展》(*De ortu et progressu chemiae*,1668)中抨击了这部著作。他提出证据认为,化学知识的肇始可以追溯到很早以前,至少可以追溯到远早于大洪水(the Deluge)②之前的塔布尔·该隐(Tubal Cain)③时期。至于赫耳墨斯,博里齐乌斯无论如何也不怀疑。不仅化学起源于埃及,医学也起源于埃及,发明这二者的人就是赫耳墨斯这位希腊的墨丘利。[8]简言之,在技艺的古老以及赫耳墨斯·特里斯麦基斯托斯在把技艺传给埃及之外的地区所起的作用问题上,康林吉乌斯是错了。

康林吉乌斯作为欧洲最博学的学者之一而受人尊重,由于博里齐乌斯的著作深深地触怒了他,他于1669年将其《论古埃及人的炼金术医学与帕拉塞尔苏斯主义者的新医学》大大扩充,出了新版,书中附有"为奥劳斯·博里齐乌斯的诽谤和敌意非难所作的辩解"(Apologeticus adversus calumnias et insectationes Olai Borrichii)一文。他论证说,要想知道最初的赫耳墨斯学说是不可能的,因为它们随着时间的推移而不断变化,我们真正知道的东西是十分可疑的。赫耳墨斯医学受到了法术和邪恶的感染……确实,埃及人的知识总的来说是邪恶的,被各种迷信仪式所充斥。那些

---

① 希腊神话中为众神传信并掌管商业、道路等的神。

② 指《旧约》第七章所说的挪亚时代的大洪水。

③ 《圣经》中所说的亚当和夏娃的长子。

论述埃及法术重要性的人本身就是可疑的，因为这种法术不是自然法术，而是超凡法术。至于帕拉塞尔苏斯及其信徒们，他们的工作已经完全腐蚀了哲学。他们的三要素毫无用处，他们的含金属药物是从以前的医生，如威兰诺瓦的阿那德（Arnald of Villanova）和拉蒙·陆里（Ramon Lull）那儿剽窃来的。其余的则是“异常地不虔不诚……异常地无根无据和荒唐”。[9]

博里齐乌斯并没有让论辩就此平息。在一部新著《奥劳斯·博里齐乌斯就赫尔曼·康林吉乌斯的指责而为赫耳墨斯主义者、埃及人和化学论者们的智慧所作的辩护》（*Hermetis, Aegyptiorum, et chemicorum sapientia ab Hermanni Conringii animadversionibus vindicata per Olaum Borrichium*，1674）中，他回答说，赫耳墨斯理所当然地是一位历史人物，嬗变技术就是他发现的。这就是埃及人能够积累足够的财富去完成他们的巨大建筑工程的原因。后来的希腊人的著作——譬如亚里士多德、泰奥弗拉斯托斯、欧几里得（Euclid）和托勒密（Ptolemy）的著作——都不像康林吉乌斯所希望的那样完美无瑕。他们受到颂扬的原因很可能归结为这样一个事实，即他们中的许多人都曾在埃及学习，并且多少都有一些在那个地方一直得到传授的古旧式智慧。

> 至于希波克拉底（*Hippocrates*）和盖伦……前者的祖国科斯（*Cos*）离埃及很近，因此这对于他获得医学知识必定极为有益；他的老师德谟克利特（*Democritus*）对

> 埃及始终很了解，无疑会启发他思考许多东西：盖伦亦曾长期生活在亚历山大，他经常建议谋求医生职业的希腊人四处旅行以获取经验。至于托勒密，他不是希腊人，而是一位亚历山大人或者培琉喜阿姆①人（*Pelusiot*），因此他是埃及人。[10]

博里齐乌斯抨击康林吉乌斯，也是由于他否定化学药物的价值。帕拉塞尔苏斯则被描绘成重新发现了早已为古埃及行家们所知晓的真理的一位大师。

由于发表在 1668 和 1674 年伦敦皇家学会《哲学学报》（*the Philosophical Transactions*）以及 1675 年发表在《博学者学报》（*the Journal des Scavans*）上的长篇评论，博里齐乌斯与康林吉乌斯之间的论战引起了人们的广泛注意。[11] 人们认为，在化学古老性问题上的争论要点，对于两部主要教科书，即康林吉乌斯的《论古埃及人的炼金术医学与帕拉塞尔苏斯主义者的新医学》和博里齐乌斯的《论化学的起源与进展》来说，是十分重要的，而这两部著作一直到 18 世纪结束都被认为是化学史学家的基本的原始资料。[12] 然而，我们要回答的问题是，为什么化学的古老性竟然会引起这么一场激烈的论战。如果我们看一看法国 17 世纪的化学教科书，我们就会发现，这些教科书的作者，譬如贝奎恩（Beguin）、勒费弗尔（Lefèvre）和莱默里（Lemery），都很少注意到这个学科的历史背景，有些作者则完全没有注意到。[13]

---

① 培琉喜阿姆（Pelusium），古埃及城市，在尼罗河最东边出口处。

然而,如果我们把目光转向这个时期的帕拉塞尔苏斯信徒和炼金术士们,我们就会看到,人们对历史的兴趣来自炼金术—医疗化学传统,而不是来自实际的教科书。彼得·塞弗里纳斯(Peter Severinus)撰写了《哲学的医学观念》(*Idea medicinae philosophicae*,1571),开始为帕拉塞尔苏斯进行辩护,书中有一章论述医术的起源和进展,其主要目的在于强调帕拉塞尔苏斯,以及在化学和医学上应用其理论的重要性。[14] 17世纪早期,达尼尔·塞纳尔(Daniel Sennert)深受医疗化学运动的影响,在其《论化学论者与亚里士多德主义者及盖伦主义者的一致与分歧》(*De chymicorum cum Aristotelicis et Galenicis consensu as dissensu*,1619)一书中,用一章的篇幅考察了化学起源。在这里,他对有些人断言亚当是第一位炼金术士的说法表示怀疑,而欣然接受塔布尔·该隐很可能是雏形炼金术的创始人的说法。[15]我们已经与塞纳尔一起涉及了传统的炼金术著作,而且只要我们考查一下两个版本的炼金术集成,我们就会看到,它们——就和约瑟夫·布莱克的演讲以及许多近代化学教科书一样——都是用历史研究开篇的。拉扎吕斯·泽兹纳(Lazarus Zetzner)《化学讲坛》(*Theatrum chemicum*,1602)的第一卷以罗贝特乌斯·瓦朗塞斯(Robertus Vallensis)的《论化学技艺的真实与古老……》(*De veritate & antiquitate artis chemicae*…)开头[16],而J. J. 曼格特(J. J. Manget)《化学珍籍文库》(*Bibliotheca Chemica Curiosa*,1702)的开头则完全是奥劳斯·博里齐乌斯的《论化学的起源与进展》的翻版[17]。

对于那些竭力要把自己的知识与大洪水之前亚当所知道的神授知识联系起来的炼金术士们来说，迷恋于自己技术的古老性是十分自然的。这些人为了使自己的作品更富有权威性，往往把自己的著作说成是某段不清楚的历史上的传奇人物所写的——这一习惯仍然是近代历史学家们一直感到头痛的事。因此，从最早的时代起，炼金术士们都力图证实他们的学问的古老性，而且正是在这个炼金术传统的基础上，才形成了标准的炼金术史和医疗化学史。这就是约瑟夫·布莱克认为"非常野蛮和荒谬的"、"越轨的"历史，但是他又在自己的演讲中将其作为早期历史的根据。[18]实际上，布莱克的历史只不过是一个悠久传统中的一章，而就实际意义来说，它则能够将这个领域的近代学术成就与若干世纪以前注重历史的炼金术士们联系起来。

如果我们再看一看博斯托克 1585 年出版的《由圣先首先传授的由统一性、和睦与协调所组成的古代医学，与来自盖伦之类的偶像崇拜者、异族人和异教徒的二元性、倾轧和对立所组成的后来的医学之间的差异》，历史与真理的关系——以及炼金术、化学与帕拉塞尔苏斯的医学化学之间的联系——就十分清楚了。[19]作为一位帕拉塞尔苏斯的信徒，博斯托克给化学下的定义与今天所使用的定义相去甚远——而且就连他对炼金术的诠释也与专门研究贱金属嬗变的技术没有多少共同之处。这本书的开头把这一点说得清清楚楚：

> 真正的和古代的医学存在于对自然之秘的探寻之中，而且要从自然本身获取。其线索要在数学资料和超自然资料二者之中查寻，之所以要在后者中查寻是因为自然是上帝的作品。这种工作是机械的，不经劳作则不能完成。它是希伯来神秘哲学（*Cabala*）的组成部分，古称圣术（*Ars sacra*）、圣识（*magna & sacra scieta*）、化学（*Chemeia*）、炼金术（*Alchimia*）或秘术（*mystica*），近来则被叫做医疗化学术（*Spagirica ars*）。[20]

因此，在博斯托克看来，真正的医学也许与我们今天叫做科学的那种东西差不多，但是，它除了最高意义上的神秘主义——炼金术之外，又什么也不是。为了与其时代一致，与炼金术传统一致，他觉得帕拉塞尔苏斯医学只要能与人类的黄金时代联系起来，就会无可争辩地被证明是真理。博斯托克为我们提供的科学史对改革表示痛惜，而把他那个世纪的成就看作对古代先知们的知识的真正复兴。那么，博斯托克的科学史就是带有目的的历史——说明炼金术—帕拉塞尔苏斯传统的来源早于它在 16 世纪晚期曾与之竞争过的盖伦—阿拉伯医学。他对他的历史的重视程度可以从他在这个问题上专门花费的篇幅来加以衡量——在一本总共只有 25 章的书中，在这个问题上所用的篇幅就达 10 章之多。[21]

显然，这些化学史和医学化学史，对于这些历史著作的作者们来说，是很重要的，因为它们把他们的科学与赫耳墨斯，偶尔也与亚当联系起来，从而提供一个环节，与上帝赋予人类的神授知识相联结。亚里士多德和盖伦的著作何以能够与这些凭证相比呢？然而，随着17世纪后期和18世纪炼金术的衰落，这个历史的原因被人们遗忘了。在波尔哈夫看来，这门科学知识简直就不需要历史，不过他觉得，“对于一个行家来说”，没有什么“比知道自己这门技艺的兴衰更有趣的了”。[22]在整个18和19世纪，人们撰写了一连串多卷本的化学史，其中许多著作现在仍然很重要，例如J. F. 格梅林(J. F. Gmelin)的3卷《化学史》(*Geschichte der Chemie*，1797—1799)[23]，本世纪还数次再版的赫尔曼·柯普(Hermann Kopp)的4卷《化学史》(*Geschichte der Chemie*，1843—1847)[24]，以及后来E. O. 冯·李普曼(E. O. von Lippmann)的《炼金术的形成与传播》(*Entstehung und Ausbreitung der Alchemie*，1919—1931)[25]。同时，M. 贝托雷[26]和威廉·奥斯特瓦尔德(Wilhelm Ostwald)[27]则在编辑各种重要的文集，著名的蒸馏器丛书的再版也在准备之中[28]。至于文献目录指南，则有约翰·弗格森(John Ferguson)的两卷《化学书目》(Bibliotheca Chemica，1906)[29]，H. C. 博尔顿(H. C. Bolton)的《化学集成》(*Bibliography of Chemistry*，1893)[30]以及杜维恩(Duveen)于1949年出版的分类目录《炼金术与化学集成》(*Bibliotheca alchemica et chemica*)[31]，其他学科则几乎没有能够与此相比的综合性书籍。

也许，这个传统之中最后一部伟大的综合性著作是詹姆斯·里迪克·帕廷顿(1886—1965)的《化学史》。他是一位相当出色的物理化学家(33岁时曾任玛丽女王学院的化学教授),早就对化学史感兴趣。他在这个领域的第一部较重要的著作是《应用化学的起源与进展》(*Origins and Development of Applied Chemistry*, 1935),其中所引用的参考文献在25000条以上[32]。随之而来的就是他那屡次再版的《化学简史》(*Short History of Chemistry*,1937)[33]和《希腊火与火药的历史》(*A History of Greek Fire and Gunpowder*, 1960)[34]。他那综括16世纪至20世纪早期的不朽之作《化学史》的第二至四卷在1960年至1964年面世——而未完成的第一卷的第一部分则在他死后的1970年出版。[35]这部著作仅篇幅而言就长达3000页左右,它必将是化学家和历史学家们的一部重要参考文献。

在帕廷顿看来,似乎无须对这门科学的历史的价值进行论证。只是在1951年英国科学史学会主席演讲中,他才离开主题特别提到,尽管理夏德·威尔施达特(Richard Willstätter)已经强调过在历史基础上进行化学教学的价值,然而这种态度似乎与新近的趋向相违背。

> 我们有些人也许忘记了,世人对这门学科的研究时时都怀有深深的敌意。我们在许多方面意识到这一点。某些中学教师的敌意是显而易见的,他们不喜欢那些涉及他们学科的历史方面的著作。在各大学,我们发现化学史科目已从学位教学大纲中消失。这件事发生的时候,人们告诉我们,是按照教化方式论述化学的时候了;科学是发展的、复杂的,学生的所有精力都

> 被吸引到掌握科学的现状上去了；论及科学的起源，不仅是浪费时间，而且只会把学生搞糊涂，只能使学生厌恶与这门学科打交道。[36]

不用说，帕廷顿对这种发展持不同意的态度。他的无机化学教科书历史色彩浓厚，他显然认为这是教授这门课程的正确方法。然而，帕廷顿对于把科学史作为一门独立学科的兴趣不大，他确实不怎么关心社会因素与化学发展的关系。在其巨著《化学史》中，他讨论了每个作者的科学著作，强调的是化学反应与化学理论，而不是这些发现与更广泛的科学史之间的关系。帕廷顿的著作尽管给人印象深刻而且很重要，但却是作为较陈旧的内在史的一座纪念碑而耸立于世的。

既然化学史具有这样一个悠久的科学的科学家—史学家传统，我们必定要问：为什么它在科学史发展中所起的整体作用这么小？1983 年 12 月出版的一期《科学史》中，J. R. 克里斯蒂(J. R. Christie)和 J. V. 戈林斯基(J. V. Golinski)在一篇关于“化学史学”的文章中写道：

> 17 和 18 世纪的化学在史学等级体系中保持着附属地位。当少数历史学家们面临她那已得到确认的颇具魅力的姊妹物理学，或者她那令人激动的姊妹生物进化论(正忙于创立)以及地质学(正忙于兴起)的竞相引诱时，她却被这些历史学家们吸引，这无疑很般配，但却显得呆滞。化学达到理智年龄，不声不响地做自己的事情，只是在后来，她与一位使她成熟起来的法国

人私通，才短暂地引起过闲言碎语。[37]

简言之，传统把化学史引向化学家，而不是引向新近形成的科学史领域。战后年代里，物理学是这门学科的中心。地质学史和进化论史最近得到发展，就与这个新近确立的领域有关系。然而，化学史是一个古老得多的研究领域，是一个在传统上被导向化学家而不是被导向历史学家的领域。它还进而在萨顿的科学等级体系中，遭受到被列在低等地位的厄运。

30 年前，在对化学史特别感兴趣的科学史学家当中，道格拉斯·麦克凯尤为突出，颇负声望。麦克凯是伦敦大学学院的科学史教授，他花费了大量精力专门研究拉瓦锡和 18 世纪的化学革命。[38]他培养了一批对类似问题感兴趣的学生，并做了许多工作使这个领域成为化学史研究中占主导地位的领域。法国的多马斯（Daumas）和美国的格拉克充实了这个领域的力量。[39]格拉克的学生玛丽·博厄斯（Marie Boas）在其《罗伯特·波义耳与 17 世纪的化学》（*Robert Boyle and Seventeenth-Century Chemistry*，1958）一书中，力图把化学史作为科学史学家们的一门基础课进行介绍。在这本书中，她把早期炼金术士们的神秘观点与波义耳“化学中的理性机械论”进行了比较，该理论实际上试图根据物体的微小粒子的大小、形状和运动来解释化学反应。尽管不否认 18 世纪的一场化学革命，但博厄斯提倡也要承认 17 世纪的一场化学革命，因为波义耳及其同行们的工作与炼金术和帕拉塞尔苏斯传统已经断裂。[40]

为了强调波义耳思想中的近代成分，玛丽·博厄斯给她《1450—1630 年的科学复兴》(*The Scientific Renaissance, 1450—1630*, 1962)一书中关于文艺复兴时期神秘的自然哲学一章，冠上了“受法术所迷”的标题。[41] 在与其丈夫 A. R. 霍尔(A. R. Hall)合作的一篇关于艾萨克·牛顿的炼金术手稿的论文中，她论证说，这些手稿是真正的化学著作——按照这个词的近代意义而言——它们不得不用一种传统的神秘语言来表达，因为当时还不存在更好的语言。[42] 甚至连赫伯特·巴特菲尔德这位在其《历史的辉格诠释》(1931)中曾经反抗过实证主义历史学家的人，也在《近代科学的起源》中写道：

> 20 世纪评论范·赫尔蒙特的人们本身就是些难以使人置信的家伙，而比较起来，在培根身上最不可思议的事情似乎也是理性主义和新式的东西。关于炼金术，发现事物的实际状态更加困难，那些专门研究这个领域的历史学家们，他们本身似乎有时也受到了上帝的天罚；因为如同撰写有关培根—莎士比亚论战或者西班牙政治的著作的那些人一样，他们似乎染上了他们开始描述的那种神经错乱。[43]

博厄斯和巴特菲尔德的评价终于受到了驳斥，这主要应归功于沃尔特·佩格尔。在近 30 年研究的基础上，他的《帕拉塞尔苏斯：文艺复兴时代的哲学医学导论》(*Paracelsus: An Introduction to Philosophical Medicine in the Era of the Renaissance*)于 1958 年面世，同年，博厄斯论述波义耳的著作面世。[44]然而，佩格尔并不像以前的某些学者们那样仅仅从范围广泛的著作中选择“近代”材料，他力图理解的是这个“完整的人”。结论是，他并不是粗俗的复兴魔法师形象，而是借助于当时流行的许多科学、医学和哲学领域进行研究的一位有影响的人物。佩格尔清楚地说明了炼金术、新柏拉图学派和诺斯替教的原始资料对于帕拉塞尔苏斯著作的影响，并指出，这些今天看起来似乎是反科学的哲学论题，当时却促使人们用新的观察方法研究自然。既依靠人类在创世中的特殊地位，又借助于相互联系的宏观世界与微观世界，帕拉塞尔苏斯坚决主张人不仅能够凭自己的力量研究自然——而且这样做去了解造物主的德性，乃是人的本分。翻遍帕拉塞尔苏斯派文献，我们看到他们经常呼吁学者们抛弃古人的著作，呼吁用新的观察和个人的经验将古人的著作取而代之。

这一切都对化学史以及科学史产生了重大影响。从佩格尔的著作中可以清楚地看到，帕拉塞尔苏斯的宇宙论和医学是与炼金术和化学理论结合在一起的。帕拉塞尔苏斯描述了一种新的神圣创世过程，根据化学分离的原理对其进行了解释，并由此导致盐、硫、汞三要素新体系。地球宇宙事件(geocosmic events)按照化学类比加以讨论，而且，人类生理学变成了化学生理学。疾病就是人体的化学机能失灵，必须用化学方法制备的药物进行治疗。帕拉塞尔苏斯的信徒们要求毁灭古代的权威，认为那些古代权威的著作应当用他们自己的“新哲学”或“化学论哲学”来取而代之。

的确,在帕拉塞尔苏斯著作中所发现的真正的化学不怎么有条理。它很少有超过更早的炼金术士们的工作的地方,甚至在其信徒们的著作中,正规地介绍化学制剂,也只是为了药用。帕拉塞尔苏斯信徒们的主要目的,始终是改革医学。然而,这也就是他们所说的化学论自然哲学。就这样,它与更多的机械论哲学相匹敌差不多达100年之久。我现在不打算详细讲述,因为这是我下一次演讲中的主题。不过我在这里要补充的是,佩格尔的著作并没有立即被化学史学家们接受。这在很大程度上是由于这样一个事实:他对于帕拉塞尔苏斯的兴趣,主要是在其医学论哲学方面。因此,他没有强调化学论哲学,这样一个帕拉塞尔苏斯学说主体中的重要组成部分。

除了深远地触及化学论哲学的影响,化学史学家们最近还对化学以一门独立学科的面貌而出现的问题进行了重新评价。这个问题与确立帕拉塞尔苏斯的化学世界观有所不同。欧文·汉纳韦(Owen Hannaway)在其《化学家与语词》(*The Chemists and the Word*,1975)一书中抓住这个问题,考查比较了帕拉塞尔苏斯的辩护士奥斯瓦德·克罗利乌斯(Oswald Crollius,1609)与安提斯·李巴尤斯(Andreas Libavius,1540—1616)①。[45]前者力图解释帕拉塞尔苏斯信徒们的化学宇宙论,把它作为他所鼓吹的化学医学的基础。后者尽管十分相信嬗变的可能性,但却反对帕拉塞尔

① 原文误为“1597—1615”。

苏斯和炼金术士们的宗教神秘主义。相反，他力图吸取早期教科书中的化学知识，并且为了教学的目的，以某种实用的方式，规定、组织这些材料。李巴尤斯的著作的确先后被简·贝吉恩(1610)[46]以及17世纪一批重要的化学教师所利用，这些教师当中的许多人准备了他们自己的教科书，教科书表明，其中绝大多数没有帕拉塞尔苏斯派宇宙论者的神秘理论。

最近，克里斯蒂和戈林斯基对与17和18世纪的化学有关的文献作了评论。实际上，他们抛弃了那种集中研究人们推翻燃素说以及承认拉瓦锡发现氧的重要性的传统内在主义方法。更确切地说，他们倾向于汉纳韦的主张，认为还须进一步着重考查李巴尤斯及其从事教学的门徒们作出的细心规定和描述。然而，我们对于17世纪化学教学的实际情况，所知道的仍然非常之少。克里斯蒂和戈林斯基举了几个重要变化的例子，这些变化的发生涉及化学理论的组织形式，他们并且强调说，需要更详尽地研究这个因素。最后，他们特别提到，最近的研究表明，18世纪的化学革命远比以前的各项研究所得出的结论要复杂得多。“现在看来”，它“似乎与气态理论、酸性理论以及新的组成概念很有关系”。[47]但是，同样重要的是这个事实，即，如同贝奎恩和17世纪教科书的作者们一样，拉瓦锡也撰写了教科书，并且与人合作研究新的命名法。“作为反对者……学生学习的新词他们同样也在学习，他们虽然不那么自觉，但却并不迟钝地认识到要抛弃旧的概念和理论。”[48]我认为这段话很清楚：就此而言，如果我们要认识化学史或科学史上已经发生过的更多的变化，我们就得把更多的时间花费在所用的教科书和教学方法上。

让我说几句离题话，谈谈何塞·玛丽亚·洛佩斯·皮涅罗(José maría López piñero)，他极为强调化学医学的引进及其与西班牙新科学的关系的重要性。他承认，

> 我们仍然远未拥有具有牢固基础的模型，去解释16和17世纪西班牙社会中科学活动的整合及其与欧洲其他地区科学发展的关系。[49]

不过，他指出了西班牙的赫尔蒙特信徒胡安·德·卡夫里亚达(Juan de Cabriada)著作的重要性，并且通过化学医生与盖伦主义者之间的激烈论战，看出了1697年塞维利亚的医学及其他科学社团创建的原因。[50]实际上，尽管洛佩斯·皮涅罗已经注意到西班牙的帕拉塞尔苏斯派教科书早在1589年就出版了，但是，西班牙医学上的传统主义者和改革者之间的论战，似乎在英国、法国和德国的论战达到高潮约100年之后，才得以发生。既然是这种情况，那么，研究近代化学的兴起及其与伊比利亚半岛变革了的医学之间的联系，便是一件相当重要的事情。只是现在，我们才开始明白，要翻印各种重要的教科书供这类研究使用。例如，洛佩斯·皮涅罗已经出版了劳伦蒂乌斯·科卡尔(Laurentius Cocar)的《揭示医学起源的对话》(*Dialogus Medicinae Fontes Indicans*, 1589)[51]，并在书中附上了注释和介绍，而唐胡安·曼努埃尔·阿雷胡拉(Don Juan Manuel Arejula)1788年写就的著作，在西班牙是一部对拉瓦锡的《命名法》加以评注的非常早的著作，这一著作已由拉蒙·加戈(Ramón Gago)和胡安·卡里略(Juan Carrillo)重新出版，书中附有详细讨论[52]。

言归正传，让我回到汉纳韦著作以及克里斯蒂和戈林斯基文章的正题上来。前者的著作集中在化学独立于炼金术和帕拉塞尔苏斯主义，而被确立为一门科学的问题上。克里斯蒂和戈林斯基提出的方法则表明，对17和18世纪的教科书提出的各种不同问题，很有可能导致对该时期的化学发展作出新的诠释。然而，二者都假定，后期的炼金术教科书和帕拉塞尔苏斯—赫尔蒙特派教科书都无须考虑。如果我们的兴趣主要是化学的确立，那么就我们所知，他们是对的。然而，在整个18世纪，一直不断地出版了许多炼金术教科书和帕拉塞尔苏斯派教科书，则是无疑的。第一张亲合力表的作者艾地安·弗朗索瓦·乔弗罗瓦(Étienne François Geoffroy)1772年在巴黎科学院宣读了一篇文章，反对增加炼金术士的数目，而蒙彼利埃(Montpellier)的化学教授加布里埃尔-弗朗索瓦·韦内尔则在狄德罗(Diderot)的《百科全书》中撰写了一篇论化学的文章，鼓吹要有一位新的帕拉塞尔苏斯出来使化学返老还童。[53]

正是在18世纪末，我们不仅看到拉瓦锡新化学的确立，而且也看到在很大程度上显示欧洲启蒙运动特征的那种严密的机械论科学遭到浪漫派的反对。这是我们目睹对炼金术和自然法术的兴趣复苏的时期。这也是我们看到梅斯梅尔(Mesmer)被控剽窃帕拉塞尔苏斯著作[54]的时期,是我们目睹哈内曼(Hahnemann)借用帕拉塞尔苏斯医学理论中的一个重要组成部分去论证其顺势疗法[55]的时期。由于这个理由,我要论证,即使文艺复兴时期化学论哲学没有形成18世纪化学主流的一个部分,但炼金术著作和帕拉塞尔苏斯派著作在当时仍然赢得了读者大众。而且,这类文献似乎形成了一个重要环节,与19世纪早期自然哲学的兴趣联系起来。

的确,对19世纪和20世纪早期的化学所进行的研究,在许多方面都还处于摇篮时代。我们固然已经有许多较早的重要研究,这包括帕廷顿《化学史》的第四卷在内,它给我们提供了近千页有关该时期的技术资料,但是最近已经有人转向,从智识史方面研究19世纪的化学。戴维·奈特(David Knight)曾经指出:

> 在[汉弗莱·]戴维[爵士]([Sir Humphry] Davy)本人及其同时代人眼中,他在19世纪早期做出的伟大成就表明,仅有力学可能还不够。化学既是一门关于诸如电之类的非物质力量的科学,又是一门关于物质的科学……按照这种分析,物质是迟钝呆惰的,而且断无真正的理由去设想存在着不同种类的物质。[56]

这种关于宇宙物质的主张与古老得多的理论非常相似。由于奈特专注于自己的问题,他描绘了“化学理论与哲学立场之间某种惊人的紧密联系,惊人的变动不居的联系”。一个中心问题是,原子的存在,以及汤姆森(Thomson)、居里

(Curie)和卢瑟福(Rutherford)20世纪早期的工作

> 似乎证实了19世纪许多伟大的化学家所相信的东西;证实了我们这个简单而和谐的世界不是由许多不同种类的原子构筑而成,而仅仅是由完全相同的粒子构筑而成的。[57]

在德国,莱因哈德·勒弗(Reinhard Löw)已经着手对化学在德国自然哲学中的地位进行研究。[58]他已证明了这样一个事实,即有机化学的发展绝没有受到这种智识运动的妨碍。因此,尽管我们都熟知李比希(Liebig)对各种自然哲学的神秘主义的抨击,但这看起来似乎不正确地牵制了我们对这场运动的研究。我想,我们可以有把握地预言,自然哲学将是一个不仅引起科学史学家们关心的领域,而且引起化学史学家们关心的领域。

对19世纪化学的研究,也在开始反映更近的史境趋向。化学社团的兴起,对于那些关心该领域的职业化的人们来说,已经成为一件重要的事情。其他人则已经对新设备和新技术的迅速发展产生了兴趣。对于因各种实际需要而在该世纪迅速发展的近代化学教育来说,自李比希开始的化学实验室的兴起,乃是必不可少的条件。在美国,尤其在政府的新赠与地,各大学总是强调化学,因为需要培养能够分析土壤样品的农业化学家。[59]

化学工业的兴起对于化学史学家来说也同样重要。这里,人们特别把兴趣集中在有机化学以及染料工业的兴起上。[60]虽然第一个用化学方法制备的染料苯胺紫是由英国的珀金(Perkin)生产的,但在商业上加以开发的却不是英国的制造商。相反,这种知识由珀金的学生们带到德国,成为19世纪后期德国化学工业的基础。这种工业在第一次世界大战爆发时被证明对德国是极有价值的。特别是1908年弗利兹·哈柏(Fritz Haber)发现氨的合成方法,迅速发展到生产阶段,这就把德国从依靠外国供应硝酸盐来生产军火的状态中解救出来。[61]哈柏亲自为政府工作,研制各种毒气,研究这些毒气在大战进攻阶段的用途。他领导的机构最后竟有150名大学教师和大约2000名助手。

当然,在和平时期,借助于肥料添加剂的发展,哈柏法应当是有益于任何地方的农民的,而且哈柏本人获得了诺贝尔奖金,但是,这个孤立的事件确实给我们提供了一个史学问题。J. R. 帕廷顿按照传统的方式讨论了哈柏法,但只是把它看作一项重要的化学发现。他对这项发现的用途则不怎么关心,因为哈柏法的应用超出了他的化学史概念的范围。我们发现我们又回到了科学史的内外关系上了。

我们可以说各种外在因素能够影响科学的发展吗?确实,就商业,譬如染料工业的例子而言,有可能训练大量的科学家去发展某个有前途的研究领域。我们也知道,19世纪后期,德国化学工业不仅通过建立工业实验室,而且通过捐款资助大学研究那些工业上特别感兴趣的科学领域,去训练这类科学家。谈到第一次世界大战,我们看到人们很快就认识到哈柏法的重要性。近来的研究进一步证实了这样的事实，即魏玛共和国[①]早在20世纪20年代就建立了一个科学机构，开始对那些被认为是有价值的研究提供资助。无疑，外在因素能将研究置于某些有希望的领域之中。

但这实质上不是20世纪的发展。弗兰西斯·培根设想的“新哲学”终将会改进各种实践方法——而法国科学院将其作为目标之一，已经为国家利益而利用了专门的科学知识。17世纪，约翰·鲁道夫·格劳伯(Johann Rudolph Glauber)也曾对经济繁荣进行预测，让农民们将丰饶时代过剩的五谷和美酒，转化为困苦时世也能够复制成啤酒、面包和酒类的浓缩物。他进而论证，认为应发展化学方法，从德国森林中获取矿石，并且通过集中努力，确定某种能够赢利的可行方法，由贱金属生产黄金。他首先就写道，需要诸如酸雾之类能够使敌兵双目失明的化学武器。这似乎很重要，他极力主张建立化学研究院，以不断研究，改进现有武器，并发明新武器。只有这样，才可能在技术上保持优先于敌方的地位。[64]

几乎没有什么疑问，包括政治、战争、商业、经济和宗教等等在内的外在因素，在科学与社会的关系方面都起着重要作用。这在很长的历史时期内能够显示出来。因此，通过财政资助开辟有希望的研究领域也是可能的，并且这样很可能会加速科学的发展。然而，诸如此类的外在因素对科学发现或主要理论的发展的影响，尚须证明。

下面，我说明一下近四分之一世纪中化学史的撰写自

---

① 1919—1933年间的德意志共和国。

身所经历的一场革命，以此结束这次演讲。起初，这门科学的历史主要是为化学家们撰写的。就这种形式而论，撰写化学史的历史悠久，这能够追溯到炼金术和帕拉塞尔苏斯派医药化学家们，他们用诸如此类的历史，将他们的知识和上帝赋予人类的原始智慧联系在一起。这样的历史被用来证明化学论哲学的古老性和权威性大大超过亚里士多德信徒和盖伦主义者。就现代形式而论——它失去了原来的论战目的——这个传统在帕廷顿的不朽著作中最为明了，这部著作是关于化学的历史事实的鸿篇概述，大部分与其他智识潮流及社会倾向没有联系。

假若我必须选择一个日期作为化学史学的转折点，我大概会选择 1958 年，因为这一年目睹了玛丽·博厄斯出版了关于罗伯特·波义耳的第一部著作，沃尔特·佩格尔出版了关于帕拉塞尔苏斯的著作。前一本书表明波义耳不仅仅是一位化学家，而且也是最有影响的机械论哲学家之一，试图以此把 17 世纪的化学放进更普遍的科学革命的画卷之中。在某种意义上人们也许会说，她是通过把化学当作物理学的一个部分而使化学具有研究价值的。佩格尔的书则不同。他评价了帕拉塞尔苏斯的整个工作，并将其医学论哲学置于该时期的智识背景之中。对我们来说，更为重要的是，他指出了化学对其医学和宇宙论思想的深刻影响。其他一些人——包括我本人——后来阐述了这个主题，以证明有过化学论自然哲学，而且许多人曾设想它将成为科学革命的“新科学”。简言之，博厄斯和佩格尔的著作——

各以自己的方式——要求化学在总体上整合进入科学革命之中。

化学作为一门独立的学科从炼金术和帕拉塞尔苏斯化学论哲学中分离出来，这是另一个问题。汉纳韦为了寻求这个问题的答案，将帕拉塞尔苏斯信徒克罗利乌斯与其对手化学家安提斯·李巴尤斯进行对比，加以考查。后者力图剥去化学的神秘外衣，并用关于化学过程的清晰定义和描述将其取而代之。尽管化学宇宙论的影响贯穿整个17世纪，但李巴尤斯的工作仍然导致出现了颇有名气的化学教科书，出现了用能使人理解的方式介绍这门学科的教学传统。最近，克里斯蒂和戈林斯基又特别提到，须对17、18世纪的化学教学传统作进一步的实证研究，并提出，如果这样做了的话，对于拉瓦锡和化学革命，我们也许会作出另一种不同诠释。

而且，我认为真的可以说，最近的研究正在改变我们对于化学史的理解。不幸的是，对于19和20世纪化学的研究，远非那么先进。这里我们只能指出，在某些研究领域，初步的研究预示着将来具有重大意义的工作。这些领域中，有一些涉及内在主义的研究。我现在正在考虑人们最近发表的关于生物化学起源的论文以及关于物理化学的详细研究工作。[65]然而，最近还有其他趋势，把这门科学与广阔得多的历史背景联系起来。我一直都对炼金术很有兴趣，炼金术延续到启蒙运动的全过程，并且似乎与浪漫主义运动和自然哲学相衔接。其他人则把19世纪的化学与智识史的其他方面联系起来。对于化学的职业化，包括化学学会的建立以及化学教育的发展在内的这个主题，有一种与日俱增的兴趣。就历史而言，从整体上考虑19世纪化学工业的兴起，并且揭示这门科学与政府的联系，也是极重要的课题。

简言之，化学史的撰写史至少已有400年，但仍然还有许多事情要做。确实，我们现在有很多理由可以说明重写内部史的必要性，但我们在智识史、政治史和社会史方面对化学和化学技术的作用仍然知之甚少。我认为，不理解化学所起的作用，就不能理解科学革命。而且我相信，在我们将来所知比现在所知更多之时，我们还会说，没有化学史知识，就不可能理解世界现代史。

注

[1] 我研究化学史在西北大学得到弗兰克·T. 古克(Flank T. Gucker)的指导，在印第安纳大学得到杰拉尔德·施米特(Gerald Schmidt)的指导。他们二人都在化学系开设了这门课程。

[2] J. R. Partington, *A Text-Kook of Inorganic Chemistry*, 6th ed. (London: Macmillan, 1950).

[3] M. Cannon Sneed and J. Lewis Maynard, *General Inorganic Chemistry*, 5th printing (New York: Van Nostrand, 1954).

[4] H. E. Roscoe and C. Schorlemmer, *A Treatise on Chemistry* (4th ed. 2 vols., 1911, 1913), 1, pp. IX-XV, "Historical Introduction".

[5] 贝托雷的历史研究为数甚多且很重要,简直无法在这里一一罗列。类似地,赫尔曼·柯普的 *Geschichte der Chemie* (4 vol., 1843—1847; Hildescheim 重印: Olms, 1966)仍然是化学史学家们所必需的东西。仅仅在知名度上略次于这些不朽之作的是拉姆塞的 *Gases of the Atmosphere*, 4th ed. (London, 1955)以及肖莱马的 *Rise and Development of Organic Chemistry*, 2nd ed. (London: Macmillan, 1894) [C. 肖莱马著,潘吉星译,《有机化学的产生和发展》,科学出版社,1978 年]。

[6] William Henry, *The Elements of Experimental Chemistry* (美国第1版根据的是伦敦第8版), 2 vol. (Philadelphia: Robert Desilver, 1819), 1, p. IX。

[7] Doulas McKie, "On Some MS. Copies of Black's Chemical Lectures — III", *Annals of Science* 16 (1960), 1-9 (1).

[8] Olaus Borrichius, *De ortu et progressu chemiae dissertatio*, *in the Biblotheca chemica curiosa*, *ed. Jean Manget*, 2 vols. (Geneva: Chouet, G. De Tournes, et al., 1702), 1, pp. 1-37 (12).

[9] Hermann Conringius, *De Hermetica Medicina Libri Duo*... 2nd ed. (Helmstedt: typis & sumptibus H. Mülleri, 1669), p. 345.

[10] 见奥劳斯·博里齐乌斯的评论, Hermetis, *Aegyptiorum*, *et chemicorum sapientia ab hermanni Conringii animadversionibus vindicata per Claum Borrichium* (Hafniae: sumptibus P. Hanboldi, 1674) in *Philosophical Transactions of the Royal Society of London* 10 (n.° 113) (April 26, 1675), 296-301 (297).

[11] 除了参考[10]以外，还见 *De ortu et progessu chemiae* 的评论，in *the Philosophical Transactions* 2 (n.° 39) (1668), 779 and the Journal des Sçavans 3 (1675), 209-11。

[12] 赫尔门·波尔哈夫在其介绍化学史的演讲中称这种交流是重要的，而博里齐乌斯的两部历史著作（包括 *De ortu et progressu chemiae*）则被曼格特列为基础炼金术集成（Bibliotheca chemica curiosa）之首。

[13] 贝奎恩和莱默里的教科书都没有从历史方面定向，而勒费弗尔则只在其 *A Compleat body of chemistry* (London: O. Pulleyn Jr, 1670)序言中第 1—4 页给出了一个简史。

[14] Petrus Severinus, *Idea Medicinae Philosophicae* (Hagae-Comitis: Adrian Clacq, 1660), pp. 1-5.

[15] Daniel Sennert, *De Chymicorum cum aristotelicis et Galenicis Consensu ac Dissensu*, 3rd ed. (Paris: Apud Societatem, 1633), pp. 17-28.

[16] Robertus Vallensis, "De veritate & antiqitate artis Chemicae & Philosophorum vel auri potabilis, testimonia & theoremata ex variis auctoribus", *Theatrum Chemicum*, ed. L. Zetzner, 6 vols. (Strassburg: Zetzner, 1659—1661), 1, pp. 7-29.

[17] 见[8]。

[18] McKie，所引书第 1 页以后。

[19] 对博斯托克著作的一般讨论，见 Allen G. Debus, "The Paracelsian Compromimse in Elizabethan England", *Ambix* 8 (1960), 71-97, 尤见 pp. 77-84。博斯托克关于原初物质的观点已由沃尔特·佩格尔作了考查，"The Prime Matter of Paracelsus", *Ambix* 9 (1961), 117-135 (124-129)。

[20] R. B. (Bostocke), Esq., *The difference between the auncient Phisicke... and the latter Phisicke* (*London: Robert Walley*, 1585), sig. B1.

[21] 论述历史的各章已由艾伦·G. 狄博斯附上介绍和注释予以重印,"An Elizabethan History of Medical Chemistry", *Annals of Science* 18 (1962), 1-29。

[22] Hermann Boerhaave, *A New Method of Chemistry; Including the Theory and Practice of that Art: Laid down on Mechanical Principles, and Accommodated to the Uses of Life. The Whole Making a Clear and Rational System of Chemical Philosophy*, trans. P. Shaw and E. Chambers (London: J. Osborn and T. Longman, 1727), p. V.

[23] Johann Friedrich Gmelin, *Geschichte der chemie seit dem Wiederaufleben des Wissenshaften bis an ende des achtzehneten Jahrhunderts*, 3 vols. (G(ttingen: J. G. Rosenbusch, 1797—1799).

[24] 见[5]。

[25] E. O. von Lippmann, *Enstehung und Ausbreitung der Alchemie*, 2 vols. (Berlin, 1919, 1931). 冯·李普曼还著有一般地论述有机化学史和科学史的著作。

[26] 例如,见其 3 卷 *Collection des anciens alchimistes Grecs* (1888; London 重印: The Holland Press, 1963),这部著作只是今天才被罗贝尔·阿勒(Robert Halleux)的研究所取代[*Les Alchimistes Grecs* (Paris: Société d'Édition "Les Belles Lettres")]。现在出了第 1 卷(内容包括莱顿纸草、斯德哥尔摩纸草和各种片断, 1981 年出版)。亦见贝托雷的 3 卷 *La chimie au moyen age* (1893; Osnabruck 重印: Otto Zeller and Amsterdam: Philo Press, 1967)。

[27] Wilhelm Ostwald, ed. *Klassiker exacten Wissenschaften*, 7 vols. (Leipzig，时间不一)。

[28] *Alembic Club Reprints*, 22 vols. (Edinburgh: E. and S. Livingstone，时间不一)。

[29] J. Ferguson, *Biliotheca chemica*, 2 vols. (Glasgow, 1906; London: Academic and Biliographical Publications, 1954).

[30] Henry Carrington Bolton, *A Select Bibliography of Chemistry* 1492—1892, (Smithsonian Miscellaneous Collections, 850, 1893; New York 重印: Kraus Reprint Corporation, 1966)。

[31] Denis I. Duveen, *Bibliotheca alchemica et chemica* (London: Dawsons, 1949).

[32] J. R. Partington, *Origins and Development of Applied Chemistry* (London: Longmans, Green & Co., 1935).

[33] J. R. Partington, *A Short History of Chemistry* (1937; 2nd edition, London: Macmillan, 1951). [J. R. 柏廷顿著，胡作玄译，《化学简史》，商务印书馆，1979 年。]

[34] J. R. Partington, *A History of Greek Fire and Gunpowder* (New York: Barnesand Noble, 1960).

[35] J. R. Partington, *A History of Chemistry* [第 1 卷第 1 部分：理论背景 (London: Macmillan; New York: t. Martin's Press, 1970)；第 2 卷：1500—1700 年(1961)；第 3 卷：1700—1800 年(1962)；第 4 卷，1800 年至现代(1964)]。

[36] J. R. Partington, "Chemistry as Rationalised Alchemy" (已故的帝国勋章获得者 J. R. 帕廷顿教授 1951 年 5 月 7 日在英国科学史学会发表的主席演讲), in *A History of Chemistry*, vol. 1, Part 2, pp. XI-XVIII (XV)。

[37] J. R. R. Christie and J. V. Golinski，“The Spreading of the Word：New Directions in the Historiography of Chemistry 1600—1800”，*History of Science* 20 (1982)，235-66 (235).

[38] 例如，见他关于拉瓦锡的两部著作：*Antoine lavoisier：The Father of Modern Chemistry* (Philadelphia：J. B. Lippincott，无日期[1935])以及 *Antoine Lavoisier：Scientist，Economist，Social Reformer* (New York：Henry Schuman，1952)。

[39] Maurice Daumas，*Lavoisier，théoricien et expérimentateur* (Paris，1955)以及许多文章，其中最重要的是亨利·格拉克在下列著作中所引用的文章：*Antoine-Laurent Lavoisier：Chemist and Revolutionary* (New York：Charles Scribner's Sons，1975). 亦见格拉克的 *Lavoisier — The Crucial Year：The Background and Origin of His First Experiments on Combustion in 1772* (Ithaca：Cornell U. P.，1961). 格拉克的论文集见 Essays and Papers in the History of Modern science (Baltimore and London：The John Hopkins U. P.，1977)。

[40] Marie Boas (Hall)，*Robert Boyle and Seventeenth-Century Chemistry* (Cambridge U. P.，1958)，pp. 231-32.

[41] Marie Boas，*The Scientific Renaissance 1450-1630* (New York：Harper，1962)，pp. 166-96.

[42] Marie Boas and A. R. Hall，“Newton's Chemical Experiments”，*Archives internationales d'histoire des sciences* 11，(1958)，113-52.

[43] Herbert Butterfield，*The Origins of Modern Science 1300-1800* (New York：Macmillan，1952)，p. 98. [赫伯特·巴特菲尔德著，张丽萍、郭贵春等译，《近代科学的起源》，北京：华夏出版社，

1988年。]

[44] Walter Pagel, *Paracelsus: An Introduction to Philosophical Medicine in the Era of the Renaissance* (Basel / New York: S. Karger, 1958). 从下列文集中发表的论文范围的广度可以清楚地看到佩格尔的影响: *Science, Medicine and Society in the Renaissance: Essays to Honor Walter Pagel*, ed. Allen G. Debus, 2 vols. (New York: Science History Publications, 1972)。

[45] Owen Hannaway, *The Chemists and the Word: The Didactic Origins of Chemistry* (Baltimore and London: The Johns Hopkins U. P., 1975).

[46] Andrew Kent and Owen Hannaway, "Some New Considerations on Beguin and Libavius", *Annals of Science* 16 (1960, 出版于1963年), 241-51。

[47] Christie and Golinski, 所引书第259页。

[48] 同上。

[49] José María López Piñero, Ciencia y Técnica en la Sociedad Española de los Siglos XVI y XVII (Barcelona: Editorial Labor, 1979), p. 12.

[50] José María López Piñero, *La Introduccíon de la Ciencia Modern en España* (Barcelona: Ediciones Ariel, 1969), p. 108-17.

[51] José María López Piñero, *Le "Dialogus"* (1589) *del Paracelsista Llorenç Coçar y la Cátedra de Medicamentos Quimicos de la Universidad de Valencia* (1591) (Valencia: Cátedra e Instituto de Historia de la Medicina, 1977).

[52] Ramon Gago and Juan L. Carrillo, *La Introduccion de la Nueva Nomenclatura Quimica y el Rechazo de la Teoria de la Acidez de*

Lavoisier en España: Edición facsímil de las Reflexiones sobre la nueva nomenclature química (Madrid, 1788) de Juan Manuel de Aréjula (Malaga: Universidad de Malaga, 1979).

[53] Allen G. Debus, "The Paracelsians in Eighteenth-Century France: A Renaissance Tration in the Age of the Enlightenment", *Ambix* 28 (1981), 36-54; Allen G. Debus, "Alchemy and Paracelsism in Early Eighteenth-Century France", 将发表于 Ingred Meerkel and Allen G. Debus eds., *Hermeticism and the Renaissance: Papers Presented at the International Conference at the Folger Library* 25-27 *March* 1982。

[54] Allen G. Debus, "History with a Purpose: The Fate of Paracelsus" (国际药学史大会开幕词,1983 年 9 月 22 日)。

[55] "同类相克"的信条对哈内曼来说是十分重要的,这也是帕拉塞尔苏斯反对盖伦医学的一句主要名言。尽管哈内曼不承认这是帕拉塞尔苏斯传统,但研究顺势疗法的早期历史学家们却都知道某些前哈内曼资料。见 A. 杰拉尔德·赫尔(A. Gerald Hull)附在托马斯·R. 埃弗雷斯特牧师(Rev. Thomas R. Everest)以下著作上的一份较早的参考文献表: *A Popular View of Homoeopathy*... (New York: William Radde, 1842), pp. 233-36。

[56] David M. Knight, *The Transcendental Part of Chemistry* (Folkestone: Dawson, 1978), p. VI.

[57] 同上,pp. i-ii。

[58] Reihnard Löw, "The Progress of Organic Chemistry During the Period of German Romantic Naturphilosophie (1795-1825)", *Ambix* 27 (1980), 1-10; Pflanzen-chemie zwischen Lavoisier und Liebig (München: Straubing, 1977).

[59] 例如，见 Roger Hahn, *The Antomy of a Scientific Institution: The Paris Academy of Science, 1666-1803* (Berkeley / Los Angeles / London: University of California Press, 1971); Jack Morrell and Arnold Thackray, *Gentlemen of Science: Early Years of the British Association for the Advancement of Science* (Oxford: Clarendon Press, 1981); J. C. Cutter, "The London Institution (1805-1933)", 莱斯特大学哲学博士论文，导师 W. H. 布罗克 (W. H. Brock); Margaret W. Rossiter, *The Emergence of Agricultural Science: Justus Liebig and the Americans* 1840-1880 (New Haven: Yale U. P., 1975)。

[60] John J. Beer, *The Emergence of the German Dye Industry* (Urbana: University of Illinois Press, 1959).

[61] K. F. Bonhoeffer, "Fritz Haber" in *Great Chemists*, Eduard Farber, ed. (New York / London: Interscience, 1961), pp. 1299-1312 (1305).

[62] J. R. Partington, *A History of Chemistry* 4, p. 636.

[63] Paul Forman, "Scientific Internationalism and the Weimar Physicists: The Ideology and Its Manipulation in Germany after World war I", *Isis* 64 (1973), 151-80; "Weimar Culture, Causality, and Quantum Theory 1918—1972: Adaptation by German Physicists to a Hostile Intellectual Environment", *Historical Studies in the Physical Sciences* 3 (1971), 1-115.

[64] Allen G., Debus, *The Chemical Philosophy: Paracelsian Science and Medicine in the Sixteenth and the Seventeenth Centuries*, 2 vols. (New York: Science History Publications, 1977), 2, pp. 425-41.

[65] 最近出版的以下著作便是一个极好的例子：Robert E. Kohler, *From Medical Chemistry to Biochemistry: The Making of a Biomedical Discipline* (Cambridge / London / New York: Cambridge U. P., 1982)。

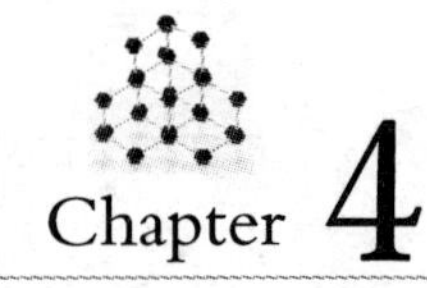

Chapter 4

第四讲

科学革命：一个化学论者的再评价

如果我们要充当负责的历史学家，我们就必须努力在历史事件发生的来龙去脉中评价它们。如果我们对于科学革命这样做，我们就发现一场主要争论集中在化学论哲学的接受与拒斥上。由于这个原因，这场争论在我们今后的科学史及医学史中必定起到十分重要的作用。

在前几次演讲中，我曾试图说明，科学革命的传统诠释为我们提供了一幅比较直截了当的画面。在这个画面中，近代科学的兴起通常被描绘成“古人”与“今人”的冲突，即那些仍然忠于亚里士多德自然哲学和盖伦医学的人与那些拥护某些以新的观察和实验为根据的“新哲学”或“新科学”的人的抗争。而正是后者，通常与17世纪的机械论哲学有联系。

这个故事的主要线索通过哥白尼、第谷·布拉赫（Tycho Brahe）、开普勒（Kepler）、伽利略以及艾萨克·牛顿爵士的工作，把我们从一幅图景引向另一幅图景。当然，还有威廉·哈维，因为他发现了血液循环。但是一般说来，相对于重述托勒密宇宙学理论以及解决与地球运转相关的运动物理学的新问题而言，生物学的发展是次要的。通常也讨论了培根和笛卡儿，但一般表明二者的科学方法论都不适宜于我们今天所知道的科学。简言之，给学生提供的是一个导致哥白尼体系和经典力学基础得以确立的科学进步的故事。

如果我们只希望沿着通向经典力学基础的阶梯前进，那么我们将看不见通向该领域的这个传统途径有什么问题。然而，如果我们打算理解那些生活在科学革命时期的学者们心目中的自然界，我们就会发现这种传统描述是不能令人满意的。如果我们试图按照他们的观点来理解他们的工作，我们在主题的选择上将无所适从，因为我们不仅必将被我们自己的问题而且也必将被他们的问题所牵制。简言之，在我们的诠释中我们必须决定我们是"现代主义的"还是"史境的"。

如果确实转向当时写出的广泛的著作，我们立即就会意识到，对新自然哲学感兴趣的作者们所关心的问题远远超出了天文学和运动物理学的范围。我认为，没有什么疑问，最热烈争论的问题与医学和化学有关——它们也许不能用我们的术语，而要用化学论自然哲学术语来定义，这种化学论哲学曾被建议用来适当地取代一直在大学里讲授的亚里士多德和盖伦的著作。甚至在谈及"古人"和"今人"时，我们都必须留心，李巴尤斯在谈及"新哲学"(*Philosophia nova*)时使用这个术语是为了刻画他所抨击的帕拉塞尔苏斯信徒的性格。此外，如果我们希望把科学革命作为一个整体重新评价的话，必须下决心研究物理学以外的其他领域。

化学史学家特别感兴趣的是瑞士—德国医生帕拉塞尔苏斯(1493—1541)及其信徒们的工作。我们今天对他们的工作感兴趣，倒不是因为他们所做的具体医学改革，而是因为他们以化学为基础所进行的新自然哲学研究。帕拉塞尔苏斯曾经受传统炼金术、医学理论和实践以及中欧采矿技术的影响，但他超越这一切形成了一个完整的自然观。他确实相信嬗变，但他的这个目标远没有医学炼金术重要。这意味着化学方法用于配制药物，但另外也意味着通向医学的神秘的炼金术途径既适用于微观现象也适用于宏观现象。

16 世纪的帕拉塞尔苏斯信徒们不同于该时期的其他自然哲学家，他们强调医学和炼金术作为重新理解宇宙之基础的重要性。帕拉塞尔苏斯信徒们的特点，是他们坚定地反对在大学里占统治地位的亚里士多德—盖伦传统。他们抵制把逻辑，因此也抵制把数学抽象作为获得真理的指南[2]，他们探寻某种东西来替换亚里士多德的元素[3]——后者的确是经院自然哲学的支柱，而他们发现盖伦主义者体液医学中的这些元素是没有价值的[4]。当然，在对经院传统的抵制上，他们强调他们探索知识的宗教性质，并且声称他们所思考的神秘真理，在后古时期的炼金术和新柏拉图主义的原本中较早就被完全认识了。帕拉塞尔苏斯信徒们确信，人类必须通过两部神学启示录圣经和创世或大自然去探寻有关造物主的知识。[5]理解后一部书，只有通过这个领域以及实验室里的新的无偏见的研究，而不是借助于有争议的这两部旧书的改写品。这样，我们就发现了早期的帕拉塞尔苏斯信徒彼得·塞弗里纳斯，他在 1571 年命令他的读者们：

> 卖掉你们的土地、房屋、衣服和珠宝，焚毁你们的书籍。另一方面，为你们自己买回结实的鞋履，到群山去旅行，探查溪谷、沙漠、海滨和最深的凹地；仔细留心

> 动物间的区别、植物的差异、各种矿石以及一切存在之物的性质和产生方式。不要对勤勉研究天文学和农民的世俗哲学感到羞臊。最后，购置煤炭，建造炉台，不厌其烦地注视并用火操作。用这种方法而不用其他方法，你们将获得对事物及其性质的认识。[6]

确实，帕拉塞尔苏斯信徒们不断倡导观察自然的新方法，而在他们看来，化学或炼金术似乎应当是这种新科学的最好范例。

帕拉塞尔苏斯信徒易于提供对《创世纪》的炼金术诠释。[7]在这里，他们把宇宙描绘成某个神圣的炼金术士把地球上的生物和物体以及天穹与未成形的原初物质（*prima materia*）分离开来的作品，差不多就像这位炼金术士能够从某种粗糙物质中提取纯的第五要素一样。在这个基础上，假定化学作为解开自然之谜的关键可能仍然具有重要性。

他们专门把注意力集中在元素的形成上，去寻找《圣经》对宇宙说明的物理真实性。[8]帕拉塞尔苏斯经常使用亚里士多德的元素，但他也引入了三要素（*tria prima*）——盐素、硫素和汞素。后者是对伊斯兰化学家们所使用的旧的金属硫—汞理论的改进，但又不同于旧的概念，它们适用于一切事物而不仅限于金属。这些要素的引入起到对古代医学和自然哲学的整体构架产生怀疑的作用，因为古代医学和自然哲学建立在亚里士多德元素的基础之上。遗憾的是，帕拉塞尔苏斯没有清楚地定义他的要素这一事实，使这个问题成为每个化学论者都可以作出各自诠释的一个很难确定的问题。

帕拉塞尔苏斯信徒们在通过化学观察或类比理解宇宙的努力中，遵循旧的但在炼金术文献中相对于制造黄金和制备药物来说又是次要的传统。[9]对于16世纪的帕拉塞尔苏斯信徒们来说，从化学上诠释宏观和微观现象是很平常的。这样，他们通过化学类比解释了世界上的气象现象。他们在地球宇宙学的水平上讨论了矿物生长和山泉起源的不同的化学诠释。[10]在寻找农业改进技术的过程中，他们把溶解盐看做是与施肥同样促使土地肥沃的重要原因。[11]这个理论竟是导致后来真正的农业化学发展的基础。

作为化学论者的帕拉塞尔苏斯信徒们竟然研究医学，这并不奇怪。他们确信，他们关于宏观世界的知识可以适用于微观世界。[12]如果空气中的硫和硝石是引起空中雷鸣电闪的原因，那么相同的臭气就可以被躯体吸入并引起高热病。体液医学被彻底否定了。这些医生不像盖伦主义者那样讨论导致疾病的体液不平衡性，而是谈论致病物质通过空气或食物进入躯体寄宿在某一器官中引起机体功能异常。这最通常是用内在生基(*archei*)——就像实验室里活生生的炼金术士一样，在各个器官中起作用的力——来解释。[13]当这些生基完全不能起作用时，也就不可能适当地排除体内的杂质，因而疾病就产生了。

文艺复兴时期产生了新的、严重的疾病，这一事实突出说明进行医学改革的必要性。冠这些疾病之首的是性病。

这些化学论医生们声称他们的新良药——通常用化学方法由金属制得——对于彻底治愈这些真正可怕的疾病是必不可少的。毫无疑问,16 世纪内服药剂的化学制备经历了重要变化。[14]帕拉塞尔苏斯本人仍然在考虑中世纪的蒸馏技术,他的制备方法的特点在于提取蒸馏的第五要素——通常证明这是新溶剂。真正的化学反应产物常常被丢弃了。但是,到进入新世纪的时候,这类实践发生了变化。医药化学论者们这时寻找新药物很少注意蒸馏的第五要素,而是更多地注意沉淀物和残余物。这对于认识化学反应来说是必要的。

尽管这种通向自然的化学途径的细节很有趣,但如果它们在当时被忽视了,就仍然不重要。事实上,1550 到 1650 年这一个世纪,帕拉塞尔苏斯信徒为一方,亚里士多德信徒和盖伦主义者为另一方的冲突是常见的。如果我们要在科学革命的解释上为他们的内容辩护,就必须确立这一点。16 世纪中期的 10 年逐渐重新找到并出版了帕拉塞尔苏斯的短文时,这些短文就开始赢得皈依新体系的人。不久出现的帕拉塞尔苏斯的不连贯论述的缩写本——例如彼得·塞弗里纳斯的缩写本——对于化学论实践者来说是必要的。

化学论医生数目的增长使这个学派成为博学的托马斯·伊拉斯都啰啰唆唆抨击的对象。伊拉斯都既在海德堡大学任神学教授又在巴塞尔大学任神学与道德哲学教授。[15]在其《关于医学新星帕拉塞尔苏斯的争论》(*Dispu-*

tationes de Medicina Nova Paracelsi，1572—1574）中，他把帕拉塞尔苏斯描绘成一个受野心和自负驱使的无知之辈、受妖魔和鬼怪点化的占星术士。他谴责帕拉塞尔苏斯是一个自相矛盾的不可信赖的庸医，一个毫无逻辑知识因此完全以混乱和莫名其妙的方式写作的人。

伊拉斯都完全反对帕拉塞尔苏斯的哲学体系，指责它把神圣的创世与化学离析相比或者从宏观和微观方面诠释宇宙，是邪恶的。伊拉斯都断言，帕拉塞尔苏斯的三要素根本不基本，同样令人讨厌。它们曾通过火法分析和蒸馏程序在热的作用下产生，因此人们仍然会确信亚里士多德的元素还是自然界的基本物质。

伊拉斯都也反对帕拉塞尔苏斯关于疾病的观点。他论证说，假定疾病是从外部进入人体的独立实体是不可思议的，并且他重申传统的体液医学确实是盖伦主义至高无上的光荣。同样令人讨厌的是帕拉塞尔苏斯在其化学制备药物方面所扮演的改革者角色。他的治疗理论已经导致使用各种矿物和金属物质——而且他还特别指出有毒的汞化合物——它们完全是致命的毒药。那么，为什么这么多人能够被帕拉塞尔苏斯完成的奇妙的治疗报告吸引到这个医学异端上去呢？伊拉斯都回答说，实际上这些“治疗”至多是暂时的。他查阅巴塞尔的档案发现，所有那些用帕拉塞尔苏斯的方法治疗的人，尽管在初期似乎有所好转，但在一年之内都死了。

在法国，化学论者与盖伦主义者之间的论战导致一系列把金属物质作为内服药使用的法律诉讼案。[16]盖伦主义者1566年取得的最初胜利证明过早，接着就是贯穿那个世纪剩下的整个时期在巴黎发生的激烈争论。在那时，约瑟夫·迪歇纳(Joseph Duchesne，1544—1609)写的一套为新化学药品辩护的丛书和小册子，引起了进一步的论战，论战的主题使欧洲各个地区的作者都极度激动。

在新世纪的早期，迪歇纳出版了两部重要著作。在书中他为化学疗法以及帕拉塞尔苏斯研究自然的态度辩护。他率直地把盖伦主义者与化学论者进行对比。他写道，前者追随盖伦，"而且似乎是根据命题只是形式上的某个王室法令，无疑地"宣布他们的医学法令。然而，化学论者们信仰的不是书本，而是理智和经验。这就是化学论医学的真正基础，就是盖伦主义者和帕拉塞尔苏斯信徒们之间分歧的根源。

尽管迪歇纳的确有时在口头上对盖伦和希波克拉底都说一些好话，但很清楚，他希望把自然哲学和医学都置于化学研究的牢固基础之上。可以理解，当时他的一些比较保守的医学同事们很可能感到受到威胁。事实确实如此，迪歇纳的书立即受到大让·里奥朗(the elder Jean Riolan)的谴责——而10年中剩下的时间里，交替对盖伦医学、帕拉塞尔苏斯医学以及化学医学进行辩护和抨击的小册子、专论、评述和书籍大量出版。到1605年，争论广为人知，超出了法国的国境，许多争论的短文被译成了他种语言。迪歇纳的著作的英译本早在1605年就出现了，而安提斯·李巴尤斯(1540—1616)则在德国出版了巴黎医学院对非难化学论者的长篇驳斥。小里奥朗(the younger Riolan)写了对这部著作的答复——这就是1607年发表的《炼金术的胜利》(*Alchymia triumphans*),它用长达九百余页的篇幅逐句答复了法国医生。

迪歇纳死于1609年,但是争论远没有结束。6年前第一个公开为他辩护的是他的朋友泰奥多尔·蒂尔凯·德·马耶尔内(Theodore Turquet de Mayerne, 1573—1655)。由于这一行动,医学界不许他的同事与他会诊,同时力劝亨利四世撤销他的各种职务。同是这一个蒂尔凯·德·马耶尔内,在巴黎被医学机构逐出成为流浪者,却于1610年应邀在英国定居,成为国王詹姆斯一世(James I)的首席御医。

在伦敦,情况大不相同。[17]医学会对一部官方药典的出版一直讨论了几十年。他们的目的,也许更多地不是为了医学同行的利益,而是为了对那些诈骗病人的药商、酿酒商以及无数没有执照的开业医生加以控制。但是,撇开伦敦学会的会员们的动机不论,很清楚,他们从一开始就乐意认可那些证明是有医学价值的化学药品。

马耶尔内在他的新家园里生活得很好。他当选为学会会员,为官方承认医疗化学药品而孜孜不倦地工作。结果1618年发布了第一部官方的国家药典。序言的作者(几乎可以肯定是马耶尔内)指出,尽管他们崇敬古人的学识并列举了他们的疗法,但"我们既不拒绝也不摒弃现代化学论者们所补充的新医术,我们在后面给了它们一席之地,为的是它们能够作为专断医学的奴仆，这样它们就会起辅助作用”。《伦敦药典》(*Pharmacopoeia Londinesis*)是折中的产物，实际上只列入很少的化学药品并引起了很大争论。这些化学药品载入这第一部官方的国家药典，很清楚，便得到了它们以前所未享有的地位——而同时，学会宣称，认可特效化学医术由它控制，通过制备方法由它决定。

如果蒂尔凯·德·马耶尔内代表极力主张采纳新医术的实用化学论者，那么罗伯特·弗拉德(1574—1637)则代表神秘主义者。[18]弗拉德与克罗利乌斯一样，可以与安提斯·李巴尤斯联系起来。1614与1615年，有几本匿名小册子被说成是由另外某个身份不明的罗齐克鲁兄弟会(Rosicrucian Brotherhood)①所写的，该会要求进行宗教改革。[19]教育改革是帕拉塞尔苏斯信徒的腔调，而且是为了与医学和化学相适应。它们很快被译成多种语言，而且引起了惊人的反响。对之进行抨击的人当中有李巴尤斯，他赞成新的化学论医学，但认为这些小册子所坚持的是他在关于奥斯瓦德·克罗利乌斯的著作中所反对的帕拉塞尔苏斯的那一套神秘主义的宇宙论。

另一方面，弗拉德希望认可以炼金术学说和基督教教义的融合为基础的神秘主义世界观。他因罗奇克鲁兄弟会的小册子而感到高兴，并于1616年写了对李巴尤斯的尖锐回答以设法取得与兄弟会的联系。[20]这本短短的《辩解书》(*Apologia*)是他在接下去的二十多年里写出的一连串出版

① 欧洲17、18世纪流行的一种秘密的宗教社团。

物中的第一本。在这些著作中，他极详细地描述了建立在宏观—微观世界的类比基础上的宇宙体系，这种类比依赖于宇宙的生命精神。对于他来说，这是真正的化学论哲学。我顺便说一句，他的著作有许多方面对于所有的科学史学家来说都是有趣的，但是我们没有时间在这里详细讨论它们。对于我们来说，重要之点在于，他对化学论哲学的神秘主义的诠释成为一场激烈争论的主题。

弗拉德有许多描述宏观世界和微观世界的书，其中的第一本发表于1617年。[21]他在这里抨击了哥白尼体系，并转而用音乐的和谐去解释太阳系。对于他来说，后者就是数学在天文学中的恰当应用。这一部分立刻引起了约翰内斯·开普勒的注意，他推迟了《世界的和谐》(*Harmonices mundi*, 1619)①一书的出版，以准备一个附录反对弗拉德的数学概念。[22]在这本著作以及第二本著作(1621)中，开普勒强调了在真正的数学家(开普勒)为一方，化学论者、炼金术士和帕拉塞尔苏斯信徒(弗拉德)为另一方，二者之间所产生的明显的差别。尽管这种差别不像开卜勒所希望的那样清楚，但数学的意义对于开普勒来说确实很不同于弗拉德。后者根据某个宇宙进程表中的某个信念寻找信条中的玄义，而前者坚决主张其假说以定量的、数学上可论证的前提为根据。如果一个假说不能容纳其观察结果，开普勒就会修改它，而弗拉德则不。

尽管弗拉德—开普勒交互攻击值得注意，但弗拉德的

---

① 其中载有行星运动第三定律。

书在法国学者中的反应是很广泛的。[23]在法国，16世纪后期和17世纪早期的帕拉塞尔苏斯医学争论，伴随着新版本和重版本化学教科书持续不断地涌现。1624年，在一位有影响的炼金术士的住宅里，当有人为14篇炼金术论文进行辩护时，新的危机发生了。根据巴黎议会的决议解散了这次集会，佐尔邦（the Sorbonne）神学院的学者们以官方的口气谴责这些论文，而在该年年底以前让·巴蒂斯特·莫兰（Jean Baptiste Morin）发表了对炼金术见解的详细的驳斥。

17世纪早期法国科学界可能再没有比马林·默森（Marin Mersenne，1588—1648）神父更有影响的人物了。作为笛卡儿、伽利略和伽桑狄（Gassendi）的朋友，他通过广泛通信联系使欧洲的学者从事最新的科学研究。1623和1625年出版的两本书清楚地表明，他感到真正以数学为基础的科学首先必须战胜化学论哲学家们的主张。他非常关心炼金术见解的神学含义，并没有完全拒斥炼金术。为了避免将来的胡思乱想，他查遍了那些控制着这个领域的国家炼金术研究机构，将它们增进人类健康和改革科学的目的作为一个整体加以考虑。对于默森来说，经过改革的炼金术将不会涉及宗教、哲学和神学问题。如果对创世进行化学诠释要得到天主教会的赞成，那么必须抵制诸如此类的梦想和思索。

默森在他的著作中指出，弗拉德是异教徒和魔术师。由于深深地受到伤害，弗拉德在两篇专论中进行了回击，他再次描述了宏观世界与微观世界的相似性、两个世界的一致性、生命精神的意义及其通过动脉系统的消散。弗拉德坚决主张，真正的炼金术以建立完整的化学论哲学为其目的，而化学论哲学是对人和宇宙二者进行解释的基础。

很清楚，弗拉德所理解的“真正的炼金术”（alchemia vera）恰恰是默森试图要回避的东西。首先，默森关于炼金术士应当割断与宗教问题的联系的警告妨碍了弗拉德。这个问题就是试图理解创世和生命精神的问题。自然界和超自然界确定是统一的——而化学对二者都起着关键作用。

默森绝望地在反对弗拉德世界体系的其他欧洲学者中寻求支持。为了这个目的，他于1628年底给他的朋友皮埃尔·伽桑狄寄去了弗拉德著作集以求帮助。当然，伽桑狄作为机械论哲学的一个创立者，由于对亚里士多德哲学的有学者风度的驳斥以及确立对物质进行原子论解释的成就，可与默森媲美。收到默森寄来的书仅两个月，伽桑狄就出了评论。伽桑狄在一段措辞激烈的话中抱怨弗拉德的观点会使“炼金术成为唯一的宗教，使炼金术士成为唯一的宗教徒，使炼金术入门成为唯一的官方许可的信条”。这是更有趣的毁灭性抨击，因为伽桑狄拒斥了威廉·哈维关于血液循环的观点。[24] 默森已经知道《心血运动论》（*De motu cordis*），他把这本书与弗拉德的书一道寄给伽桑狄，确信此书的作者就是他敌手的门徒。伽桑狄正确区分了他的实验工作和其他人的神秘的“解剖学论文”。弗拉德以前就记述了人体通过动脉系统进行的神秘的生命精神循环，但伽桑狄拒斥了哈维和弗拉德，因为他坚持盖伦的血液流动体系。

这是第一场关于哈维循环的重要论战，而有趣的是，在一部集中论述用化学方法研究自然之缺陷的书中，哈维被说成是罗伯特·弗拉德的门徒。

默森在反对弗拉德运动中所接触的人当中，让·巴蒂斯特·范·赫尔蒙特(1579—1644)是继续与法国学者保持频繁通信联系的人。[25]在最早的一封信(1630)中，范·赫尔蒙特回答了对有关伽桑狄当时回击弗拉德的作用的怀疑。这位比利时医生和化学论者谈到拙劣的医生和糟糕的炼金术士"动摇不定的弗拉德"时毫不含糊，认为伽桑狄不应当在一个学识浅薄的人身上浪费时间。

然而，如果范·赫尔蒙特至少暂时包括在默森的圈子里，他的工作就会暴露出他曾深受早期帕拉塞尔苏斯信徒的影响。在他的第一个出版物(1621年)中，他吹捧过宏观世界与微观世界的相似性、帕拉塞尔苏斯及其三要素，还曾把巫术说成是"最深奥的先天知识"。[26]这本书导致西班牙宗教法庭对他的异端邪说提出起诉，随之把他软禁起来。还有人控告他追随帕拉塞尔苏斯及其门徒宣扬其化学论哲学，他一直在全世界散布的不只是"永恒的黑暗"。[27]

范·赫尔蒙特后来的著作显示出更多的批判态度，但他与其前辈仍有许多联系。我们始终看得到他对改革的强烈要求。"必须摧毁整个古代的自然哲学，并且形成新自然哲学学派的学说。"[28]古代的科学和医学表征为数学的和逻辑的，必须不惜任何代价避免这样，以支持对自然界的真正观察的态度。范·赫尔蒙特提出，新哲学应当努力拒斥主要通过数学去诠释自然的任何概念。[29]

人们可能注意到，范·赫尔蒙特的整个工作都与自然和宗教的紧密联系有关。他要我们首先再看一看《创世纪》中创世的说明。这样——正如弗拉德那样——就可以弄清创世的次序以及水和气这两种真正的元素。三要素对实用化学家也许是有用的，但却不是基本的。[30]自然界的答案要在新的观察结果中发现——而为我们提供寻找真理的最大机会的是化学。

> 我赞美上帝，他曾呼唤我脱离其他职业去研究火的技术。对于真正的化学来说，其原理不是通过清谈获得，而是自然知晓，借助火弄明：它训练洞察自然奥秘的理解力，并且比其他所有的学科加起来更能导致对自然的进一步研究：它甚至洞察到真正的最深奥的真理。[31]

由此联想到的是对量化的理解——这里把它理解为实验室里的称重和测量而不是数学抽象——这样也许能提供新的见识。[32]尽管柳树试验是最闻名的例子，人们也可以引证他在表现比重以及精确的实验室温标方面的影响。他在工作过程中曾坚持化学变化过程中物质的守恒性和重量的稳定性。

范·赫尔蒙特的医学反映了他的背景情况。[33]由于不愿意接受古代医学的论题，他也曾受到那些愿意接受据说是帕拉塞尔苏斯所说的一切人的干扰。这样，在这些后期著作中，范·赫尔蒙特拒斥把人假定为更大世界之更精密的复制品的微观世界学说。[34]这仍然不妨碍他呼吁注意在人和自然中发现的许多大体上的相似性。作为一个例子，他研究了人体中疾病的地域性现象，尽管这种现象复杂得多，但与地球中金属的生长在许多方面是相似的。范·赫尔蒙特与弗拉德一样关心生命精神。[35]他相信血液中存在生命精神，是他坚决拒斥血液流动的重要因素。同样重要的是他对消化的化学研究，这项研究导致他用生理学的和无机的样本描述酸碱中和作用。[36]

与其他化学论哲学家一样，范·赫尔蒙特也追求教育改革。我们在前面曾提到这种需要，但每次提出并认真研究许多教育改革计划时，人们就把范·赫尔蒙特的著作印出来。确信研究自然的目的仅仅是为了医学进步——以及学习有关我们的造物主的有关知识——他痛苦地记述了他所受的教育没有导致可靠的知识并且曾使他一度谢绝硕士学位。[37]他论证说，如果要进步，我们就必须拒斥亚里士多德式的大学研究，而把新科学建立在全新的教育体系上。学生们的学习应当从简单的科目开始：算术、几何学、地理学、各民族习俗以及对动物和植物的观察。这样的基础学习三年之后，青年人也许就可以着手他们的教育的重要部分，即研究自然。但是，对于范·赫尔蒙特来说，研究自然的实施，也许就是用唯一的一种方法才能令人满意，即通过化学考试。这些研究进行四年，其完成

> 不是借助于直率的讨论叙述，而是通过火的手艺示范。事实上，自然就是通过蒸馏、润湿、干燥、煅烧和

> 分解来调节她的运作的，而[化学]玻璃的制作正是用这些相同的方法完成这些相同的操作的。因此，工匠通过改变自然的操作获得同样的性质和知识。不论哲学家可能具有的才智和敏锐的判断力多么自然，没有火，他仍然不会触及自然事物的根源或基本知识。因此，每个人都会被上千个不借助火就弄不清的思想或疑虑所欺骗。所以，我承认，没有什么东西能比火更彻底地使渴求知识的人认识一切可知的事物。[38]

确认新的化学论哲学必须取代现在已经死去的学院研究，范·赫尔蒙特断言，看看受过他所提出的那种教育的学生“超过大学哲学家们以及学院的愚蠢推理”多少，将会是一个小小的奇迹。

范·赫尔蒙特号召对教育进行化学论改革，在英国得到了极强烈的响应。诺亚·比格斯(Noah Biggs)于1651年要求进行显然是以赫尔蒙特的蓝图为基础的医学和自然哲学课程改革，同时，约翰·弗伦奇(John French)记述了“大海彼岸一所衰落的著名大学”通过对化学的赞助已经恢复到它以前的全盛时期。更重要的是牛津和剑桥在1654年进行了一场有关教育的争论。[39]是年，新教牧师、外科医师和化学论者约翰·韦伯斯特(John Webster)要求废除这两个学府的学究方法。他论证说，神圣的真理只会通过上帝在《圣经》中给人的启示以及他所创造的自然的研究获得。为了展示医学和科学中新近研究的广泛知识，他呼吁把重点放在数学中的新技能、哥白尼体系和哈维循环的教授以及原子论解释上。但最重要的是,他感到正是帕拉塞尔苏斯学派和赫尔蒙特学派的化学论者们才真正提供了可被其他人竭力仿效的观察和实验纲领。根本的关键是化学,因为化学使用手工操作法教人们阐明自然的奥秘。像范·赫尔蒙特一样,韦伯斯特注意到哲学家没有火的知识就不能研究科学的根本。他进而讲,在化学实验室里工作一年,会比对亚里士多德的原文争论几世纪更有益。

对于这种进步,我们也许要从这种新科学、这种化学论科学研究起,这恰恰是经院体系似乎必然毁灭的原因。当塞思·沃德(Seth Ward)和约翰·威尔金斯(John Wilkins)起来直接对韦伯斯特支持的新课程提出非难[40]时,他们发现他们自己的建立在机械论原理基础上的新科学梦想,与"实验"化学论者相比,和盘踞在大学里的亚里士多德学派有更多的相同之处。无疑,这些化学论者是以对自然进行观察研究和实验研究的纲领为基础的教育改革的最畅言无忌的支持者。

当然我们都知道,无论是牛津还是剑桥——更不用说欧洲的其他大学——都没有重新组织它们的课程以适应化学论者们改革的要求。获胜的是机械论者的"新哲学"而不是化学论者们的"新哲学"。这也许是由于他们的科学有优越性——或者是由于——至少部分地是——因为雅各布对英国的建议,那些把化学论哲学与宗教狂以及导致国内战争的激进政治相联系的信仰自由的教徒拒斥化学论哲学。[41]我们的确知道,很少有炼金术士或者赫尔蒙特派化学论哲学家在 1662 年伦敦的皇家学会建立时成为其会员。我们也没见到他们被大量地吸收进 17 世纪后期或 18 世纪的其他国家的科学团体。我们从他们的许多出版物中知道，他们仍然在活动，但是他们没有被吸收进学会清楚地表明他们并不是新的科学大家庭中的成员。

然而，这并不意味着他们在伦敦和巴黎的科学学会建立以前的世纪里不重要。在那个时期有化学论哲学家们撰著的大量文献——而且我们已经指出过，一个紧密联系在一起的欧洲学者圈子里曾广泛阅读了这些著作。由化学论者们的出版物完全可以知道，这些人中有伊拉斯都、开普勒、李巴尤斯、默森、伽桑狄以及许多次要的人。他们与化学论者们的争论涉及与新科学建立有关的许多关键问题：古代哲学家的价值，新的观察和实验的作用，涉及推理、运动研究和实验室时数学在诠释自然中的用途。哈维关于血液循环的观点是在这种前后关系中首先争论的，而且这里不亚于哥白尼学派的争论，科学与宗教的关系是人们关心的重要主题。这些主题以及其他主题必定是所有关心这个时期的科学革命的人都感兴趣的。

理解化学论哲学的重要性是困难的，如果我们把自己仅仅局限在医学范围内：化学制备药物的引入、疾病的概念乃至医学理论的一般问题。帕拉塞尔苏斯信徒不仅仅是医生。还必须把他们理解为化学论自然哲学家、自觉探索用建立在真正的宗教基础上的观察和新的化学基础上的观察这两根支柱上的"新哲学"去置换古代知识的学者。由于大小世界的相似性——不论在不加夸张地考察帕拉塞尔苏斯学派，还是隐喻地考察成熟的范·赫尔蒙特——医学在他们关于自然的新知识概念中都起着中心作用。

人们可能注意到这些化学论哲学家们在17世纪期间普遍接受化学医术，在持续地对生命精神进行化学研究以及在广泛支持由一些作者诸如威利斯或德拉博·西尔维乌斯(de la Böe Sylvius)提出的着重化学的生理学体系方面的直接影响。确实，没有这种广泛公认的对自然界的态度，理解查尔顿(Charleton)、波义耳、梅奥(Mayow)、贝歇尔(Becher)、格利森(Glisson)甚或艾萨克·牛顿都很困难。[42]医学课题经常得到讨论并且与元素理论、化学理论、地球的性质以及金属的产生和生长联系起来理解。学习这种化学论哲学的学生不仅必须准备研究这些课题，还应当要求在化学论教科书中找到与教育改革、经济改革及农业改革一样广泛的另外的课题。

有一些人会争辩说我们忽视这些著作也许不会错。我在前面一个演讲中曾谈到玛丽·赫斯捍卫对科学革命的更传统的解释。她拒斥近来关于赫尔蒙特原始资料的研究，并且说，“把更多的光投射在画面上会歪曲已经看见的东西”。[43]我们可以反问我们看见过的画面是否被歪曲了，并且建议增加光把它弄清楚。的确，对17世纪的帕拉塞尔苏斯—赫尔蒙特学派的大量文献的研究，会不可估量地增加我们对近代科学兴起的理解的复杂性。但是如果当时阅读并思考了这类文献，我们能够忽视它吗？假定我们对它的无知会保证更准确的诠释，这对我来说是一个极为奇怪的结论。应当把传统的描述称为对16、17世纪物理学史和天文学史的描述。

如果我们要充当负责的历史学家，我们就必须努力在历史事件发生的来龙去脉中评价它们。如果我们对于科学革命这样做，我们就会发现一场主要争论集中在化学论哲学的接受与拒斥上。由于这个原因，这场争论在我们今后的科学史及医学史中必定起到十分重要的作用。

注

[1] 这个演讲大量吸取了我过去二十多年的许多研究结果。我特别提请读者参考"The Chemical Philosophy: Chemical Medicine from Paracelsus to van Helmont", *History of Science*, 12(1974), 236-59。此文根据 *The Chemical Philosophy: Paracelsian Science and Medicine in the Sixteenth and Seventeenth Centuries* (2 Vol., New York: Science History Publications)部分手稿写成，该书直到1977年才出版。我对于化学论哲学与科学革命其他方面的关系的看法，在我的 *Man and Nature in the Renaissance* (Cambridge et al: Cambridge U. P., 1978)[艾伦·G. 狄博斯著，周雁翎译，《文艺复兴时期的人与自然》，上海：复旦大学出版社，2000年]中进行了讨论。

[2] 见 Allen G. Debus, "Mathematics and Nature in the Chemical Texts

of the Renaissance, *Ambix*, 15(1968), 1-28, 211.

[3] 关于亚里士多德的元素和帕拉塞尔苏斯的要素,见 Walter Pagel, *Paracelsus: An Introduction to Philosophical Medicine in the Era of the Renaissance* (Basel / New York: Karger, 1958), pp. 82-104。

[4] Debus, *The Chemical Philosophy* 1, 58-59.

[5] 同上,pp. 69-70。

[6] Petrus Severinus, *Idea medicinae philosophicae* (1571; 3rd ed., Hagae Comitis: 1660), p. 39.

[7] 同上, pp. 55-56 及其他各处。

[8] 同上, pp. 57-58,亦见[3]。

[9]Allen G. Debus,"The Pharmaceutical Revolution of the Renaissance", *Clio medica*, 11(1976), 307-17.

[10] Allen G. Debus, "Edward Jorden and the Fermentation of the Metals: An Iatrochemical Study of Terrestrial Phenomena" in *Toward a History of Geology: Proceeding of the New Hampshire Inter-Disciplinary Conference on the History of Geology, September 7-12, 1967*, Cecil J. Schneer, ed. (Cambridge, Mass: M. I. T. Press, 1969), pp. 100-21.

[11] Allen G. Debus, "Palissy, Plat and English Agricultural Chemistry in the 16th and 17th Centuries", *Archives Internationales d'Histoire des Sciences* 21 (1968), 67-88.

[12] Debus, *The Chemical Philosophy* 1, 96-109.

[13] Walter Pagel, *Paracelsus*, pp. 104-12.

[14] 特别使人感兴趣的研究包括 Robert P. Multhauf, "Medical Chemistry and 'The Paracelsians'", *Bulletin of the History of Medicine* 28(1954) 101-26 以及"The Significance of Distillation in Renaissance Medical Chemistry", *Bulletin of the History of Medi-*

cine30（1956）329-46；Wolfgang，Schneider，"Der Wandel des Arzneishatzes im 17. Jahrhundert und Paracelsus"，*Sudhoffs Archiv für Geschichte der Medizin und der Naturwissienschaften* 45（1961），201-15；*geschichte der pharmazeutische Chemie*（Weinheim：Verlag Chemie，1972），亦见该丛书第1、5和14卷，*Veröffentlichungen aus dem Pharmaziegeschichtlichen Seminar der Technischen Universität Braunschweig*（Prof. Dr. Wolfgang Schneider）：G. Schröder，*Die pharmazeutisch-chemischen Produckte deutscher Apotheken im Zeitalter der chemiatrie*（Bremen，1957），H. Wietschoreck，*Die pharmazeutisch-chemischen Produckte deutscher Apotheken im Zeitalter der Nachchemiatrie*（Braunschweig，1962），M. Klutz，*Die Rezepte* in *Oswald Crolls Basilica chymica*（1609）*und ihre Beziehungen zu Paracelsus*（Braunschweig，1974）。

[15] Debus，*The Chemical Philosophy* 1，第131—134页讨论了伊拉斯都的工作。

[16] 关于16世纪后期和17世纪早期法国的争论，见上书，第145—73页。

[17] 对导致《伦敦药典》出版的英国情况，上书第173—91页作了概括。乔治·厄丹(Georg Urdang)对该药典的出版作了许多基本研究。这在他的 *Pharmacopeia*：*Pharmacopeia Londinensis of 1618 Reproduced in Facsimile with a Historical Introduction by Georger Urdang*（Madison：University of Wisconsin Press，1944）再版本前言中作了概括。

[18] 我关于罗伯特·弗拉德的工作及其与化学论哲学关系的观点在许多文章中可以找到。Chemical Philosopgy 1 第205—93页中概括了许多这类补充材料。

[19] 关于弗拉德与罗克奇鲁文献的讨论，见上书第 208—24 页。亦见 *Robert Fludd and His Philosophical Key: Being a Transcription of the Manuscript at Trinity College, Cambridge*, with an Introduction by Allen G. Debus (New York: Science History Publication, 1979)。

[20] Robert Fludd, *Apologia compendiaria frater nitatem de Rosea Cruce suspicionis et infamiae maculis aspersam, veritatis quasi fluctibus abluens et abstergens* (Leiden: Godfrid Basson, 1616)次年大大扩充为 Tractatus apologeticus interitatem societatis de Rosea Cruce defendens (Leiden: Godfrid Basson, 1617)

[21] Robert Fludd, *Utrisque cosmi maioris scilicet et minoris metaphysica, physica at que technica* (Oppenheim: T. De Bry, 1617).

[22] Debus, *The Chemical Philosophy* 1, 256-60.

[23] 法国有关弗拉德与默森和伽桑狄争论的情况，在上书第 260—79 页中作了讨论。

[24] Allen G. Debus, "Robert Fludd and the Circulation of the Blood", *Journal of the History of Medicine and Allied Science* 16 (1961), 374-93; "Harvey and Fludd: Irrational Factor in the Rational Science of the Seventeenth Century", *Journal of the History of Biology* 3 (1970), 81-105.

[25] 关于范·赫尔蒙特，见 Debus, *The Chemical Philosophy* 2, 295-379，关于范·赫尔蒙特的医学观点，见 Walter Pagel, *Joan Baptista van Helmont: Reformer of science and medicine* (Cambridge et al.: Cambridge U. P., 1982)。

[26] J. B. van Helmont, *Disputatio de magnetica vulnerum naturali et legitima curatione, contra R. P. Joannem Roberti* (Paris: Victor Leroy, 1621). 一部重要著作。这清楚表明他当时完全不同于他

在 1648 年以后发表的专论所表明的那个人。

[27] 关于剥夺他的公权和逮捕他的情况，见 Pagel，*van Helmont*，pp. 10-13。

[28] J. B. van Helmont，*Ortus medicinae. Id est，initia physicae inaudita. Progressus medicinae novus，in morborum ultionem，ad vitam Iongam*（Amsterdam：Ludovicus Elsevir，1648；布鲁塞尔重印本：Culture et civilisation，1966），p. 6（摘自"Promissa authoris"）。

[29] Debus，*The Chemical Philosophy* 2，112-17.

[30] 同上，pp. 317-22。

[31] van Helmont，Ortus，p. 463[摘自"Pharmacopolium ac dispensatorium modernorum"(Sect. 32)]. 我在这里引用的是约翰·钱德勒(John chandler)的现代英译本[*Oriatrike or Physick Refined...*（London：Lodowick Loyd，1662），p. 462]。

[32] Debus，*The Chemical Philosophy* 2，pp. 327-29.

[33] 关于范·赫尔蒙特的生物学思想和疾病理论的最完整的讨论，可以在 Walter Pagel，*van Helmont*，pp. 96-198 中找到。

[34] van Helmont，*Ortus*，p. 328.

[35] Allen G. Debus，"Chemistry and the Quest for a Material Spirit of Life in the Seventeenth Century"，in *IV Colloquio Internazionale del Lessico Intellecttuale Europeo*（Rome，7-9，January 1983）(印刷中).

[36] Pagel，*van Helmont*，pp. 129-40.

[37] 范·赫尔蒙特的自传"Studia authoris"，刊于各个版本的 *Oritus* 和 *Opera* 之中。

[38] van Helmont，*Ortus*，pp. 49-50[摘自"Physica Aristotelis at Galeni ignara"(sects. 9-11)]。我在这里引用的是钱德勒的译本（见注

[31]) p. 45。

[39] John Webster, *Acadeiarum Examen, or the Examination of Academies...* (London: Giles Calvert, 1654)重刊于 Allen G. Debus, *Science and Education in the Seventeenth Century: The Webster-Ward Debate* (London: Macdonald and New York: American Elservier, 1970), pp. 67-192.

[40] 重刊于 Debus, *Science and Education*, pp. 193-259。

[41] Margaret D. Jacob, *The Newtonians and the English Revolution* 1689—1720 (Ithaca: cornell U. P., 1976).

[42] Debus, *The Chemical Philosophy* 2, pp. 447-537 中较详细地指出了这些联系。

[43] Mary Hesse, "Reasons and Evalution in the History of Science", in *Changing Perspectives in the History of Science: Essays in Honour of Joseph Needham*, Mikulás Teich and Robert Young eds. (London: Heinemann, 1973), pp. 127-47 (143).

西汉人名对照表

Adam　亚当

Albert Magenus　艾伯塔斯·马格内斯

Alexander of Tralles　查理斯的亚历克山大

Archimedes　阿基米得

Arejula, Don Juan Manuel　唐胡安·曼努埃尔·阿雷胡拉

Aristotle　亚里士多德

Arnald of Villanova　威兰诺瓦的阿拉德

Avicenna　阿维森纳

Bacon, Francis　弗兰西斯·培根

Bacon, Roger　罗杰·培根

Bailly, Jean Sylvain　让·西尔万·巴伊

Beguin, Jean　让·贝奎恩

Becher, J. J.　贝歇尔

Berthelot, M.　贝托雷

Biggs, Noah　诺亚·比格斯

Black，Joseph　约瑟夫·布莱克

Boerhaave，Hermann　赫尔曼·波尔哈夫

Bolton，H. C.　博尔顿

Bostocke，R.　博斯托克

Borrichius，Olaus　奥劳斯·博里齐乌斯

Boyle，Robert　罗伯特·波义耳

Brahe，Tycho　第谷·布拉赫

Brock，W. H.　布罗克

Bruno，Giordano　乔达诺·布鲁诺

Butterfield，Herbert　赫伯特·巴特菲尔德

Cabriada，Juan de　胡安·德·卡夫里亚达

Calvin，John　约翰·加尔文

Cantor，Moritz　莫里兹·康托

Carrillo，Juan　胡安·卡里略

Cassirer，Ernst　恩斯特·卡西尔

Cavendish，Henry　亨利·卡文迪什

Celsus　塞尔苏斯

Charleton，Walter　沃尔特·查尔顿

Christie，J. R. R.　克里斯蒂

Clagett，Marshall　马歇尔·克拉杰特

Cocar，Laurentius　劳伦蒂乌斯·科卡尔

Cohn，I. Bernard　I. 伯纳德·科恩

Comte，August　奥古斯特·孔德

Conant, James B. 詹姆斯·B. 科南特
Conringius, Nicholas 赫尔曼·康林吉乌斯
Copernicus, Nicholas 尼古拉·哥白尼
Cornell, Ezra 埃兹拉·康奈尔
Crollius, Oswald 奥斯瓦德·克罗利乌斯
Curie, Marie and Pierre 玛丽·居里和皮埃尔·居里
Cuvier, George 乔治·居维叶

Darwin, Charles 查尔斯·达尔文
Daumas, M. 多马斯
Davy, Sir Humphry 汉弗莱·戴维爵士
Debus, Allen G. 艾伦·G. 狄博斯
Dee, John 约翰·迪伊
Democritus 德谟克里特
Descartes 笛卡儿
Diderot, Denis 德尼·狄德罗
Dobbs, B. J. T. 多布斯
Draper, John William 约翰·威廉·德雷珀
Duchesne, Joseph 约瑟夫·迪歇纳
Duhem, Pierre 比埃尔·迪昂
Duveen, Denis I. 丹尼尔·I. 杜维恩

Erastus, Thomas 托马斯·伊拉斯都
Euclid 欧几里得

Everest, Thomas R. 托马斯·R. 埃弗雷斯特

Ferguson, John 约翰·弗格森
Fleming, Donald 唐纳德·弗莱明
Fludd, Robert 罗伯特·弗拉德
Freind, John 约翰·弗赖恩德
French, John 约翰·弗伦奇

Gago, Ramón 拉蒙·加戈
Galen 盖伦
Galileo 伽利略
Gassendi, Pierre 皮埃尔·伽桑狄
Gelbart, Nina 尼娜·盖尔巴特
Geoffroy, Étienne François 艾地安·弗朗索瓦·乔弗罗瓦
Gillispie, Charles C. 查尔斯·C. 吉利斯皮
Glauber, Johnn Rudolph 约翰·鲁道夫·格劳伯
Glisson, Francis 弗兰西斯·格利森
Gmelin, J. F. 格梅林
Golinski, J. V. 戈林斯基
Gucker, Frank T. 弗兰克·T. 古克
Guerlac, Henry 亨利·格拉克
Gui de Chauliac 居伊·德·肖利亚克
Guinter of Andernach 安德纳森的居恩特

Haber, Fritz 弗利兹·哈柏

Hahnemann, Samuel 塞缪尔·哈尔内

Hall, A. R. 霍尔

Hall, Marie Boas 玛丽·博厄斯·霍尔

Halleux, Robert 罗贝尔·阿勒

Hannaway, Owen 欧文·汉纳维

Harvey, William 威廉·哈维

Heath, Sir Thomas Little 托马斯·利特尔·希思爵士

Heiberg, John Ludvig 约翰·路德维西·海伯格

Henry IV 亨利四世

Henry, William 威廉·亨利

Hermes Trismegistus 赫尔默斯·特里斯麦基斯托斯

Hesse, Mary 玛丽·赫斯

Hill, Christopher 克里斯托弗·希尔

Hippocrates 希波克拉底

Hobbs, Thomas 托马斯·霍布斯

Homer 荷马

Hull, A. Gerald A. 杰拉尔德·赫尔

Huxley, T. H. 赫胥黎

Jacob, Margaret 玛格丽特·雅各布

James I 詹姆斯一世

Kekule, F. A. 凯库勒

Kepler, Johannes 约翰内斯·开卜勒

Knight, David 戴维·赖特

Kopp, Hermann 赫尔曼·柯普

Koyré, Alexandre 亚历山大·柯瓦雷

Kuhn, Thomas S. 托马斯·S. 库恩

Laplace, Pierre Simon de 比埃尔·西蒙·德·拉普拉斯

Lavoisier, A. L. 拉瓦锡

Lecky, W. E. H. 莱基

Lefèvre, N. 勒费弗尔

Lemery, N. 莱默里

Liebig, Justus 尤斯图斯·李比希

Libavius, Andreas 安提斯·李巴尤斯

Lippmann, E. O. von 冯·李普曼

Löw, Reinhard 莱因哈德·勒弗

López Piñero, José María 何塞·玛丽亚·洛佩斯·皮涅罗

Lull, Ramon 拉蒙·陆里

Luther, Martin 马丁·路德

McKie, Douglas 道格拉斯·麦克凯

Manget, J. J. 曼格特

Mayerne, Theodore Turquet de 泰奥多尔·蒂尔凯·德·马耶尔内

Maynard，J. Lewis 路易斯·J. 梅纳德
Mayow，John 约翰·梅奥
Melanchthon，Philipp 菲利普·梅兰西顿
Merkel，Ingrid 英格里德·默克尔
Mersenne，Marin 马林·默森
Merton，Robert King 罗伯特·金·默顿
Merz 梅尔茨
Mesmer，Franz A. 弗朗兹·A. 梅斯梅尔
Montucla，J. E. 蒙丢克拉
Morin，Jean Baptiste 让·巴蒂斯特·莫兰
Murray，John J. 约翰·J. 默里

Needham，Joseph 李约瑟
Neuberger，Max 马克斯·纽伯格
Neugebauer，Otto 奥托·纽格鲍尔
Newton，Isaac 艾萨克·牛顿

Oribasius 奥利巴苏斯
Ostwald，Wilhelm 威廉·奥斯特瓦尔德

Pagel，Julius 朱利叶斯·佩格尔
Pagel，Walter 沃尔特·佩格尔
Paracelsus 帕拉塞尔苏斯
Partington，James R. 詹姆斯·R. 帕廷顿

Paul of Aegina　埃伊纳岛的保罗

Perkin, William　威廉·珀金

Plato　柏拉图

Priestley, Joseph　约瑟夫·普里斯特利

Proclus　普罗克拉斯

Ptolemy　托勒密

Ramsay, Sir William　威廉·拉姆塞爵士

Rattansi, P. M.　拉坦西

Riolan, Jean　让·里奥朗

Roscoe, H. E.　罗斯科

Rutherford　卢瑟福

Sarton, George　乔治·萨顿

Schmmidt, Gerald　杰拉尔德·施米特

Schorlemmer, C.　肖莱马

Sennert, Daniel　达尼尔·赛纳尔

Servetus, Michael　迈克尔·塞尔维特

Serverinus, P.　塞弗里纳斯

Shakespeare, William　威廉·莎士比亚

Sigerist, Henry　亨利·西格里斯特

Smith, Cyril S.　西里尔·S. 史密斯

Sneed, M. Cannon　M. 坎农·斯尼德

Sudhoff, Karl　卡尔·萨德霍夫

Sylvius, Franciscus de la Boë　弗兰西斯科·德·拉·博·

西尔维乌斯

Tannery, Paul 保罗·坦纳雷
Theophrastus 泰奥弗拉斯托斯
Thomas, Keith 基恩·托马斯
Thomson, J. J. 汤姆森
Thorndike, Lynn 伦恩·桑代克
Tubal Cain 塔布尔·该隐

Urdang, Georg 乔治·厄丹

Vallensis, Robertus 罗贝特乌斯·瓦朗塞斯
van Helmont, J. B. 范·赫尔蒙特
Venel, Gabriel-François 加布里埃尔-弗朗索瓦·韦内尔
Vesalius, Andreas 安德烈·维萨留斯

Walsh, James 詹姆斯·沃尔什
Ward, Seth 塞斯·沃德
Webster, John 约翰·韦伯斯特
Westfall, R. S. 韦斯特福尔
Whewell, William 威廉·惠威尔
White, Andrew D. 安德鲁·D. 怀特
Wilberforce, Samuel 塞缪尔·威尔伯福斯
Wilkins, John 约翰·威尔金斯

Willis, Thomas　托马斯·威利斯

Willstätter, Richard　理夏德·威尔施达特

Wimpenaeus, Johann A.　约翰·A. 威姆帕纽斯

Yates, Frances A.　弗朗西斯·A. 耶茨

Zetzner, Lazarus　拉札吕斯·泽兹纳

Zwingli, Ulrich　乌尔里西·茨温利

ALLEN G. DEBUS

SCIENCE AND HISTORY
A Chemist's Appraisal

1984

LECTURES GIVEN AT THE UNIVERSITY OF COIMBRA 1983

by

Allen G. Debus

The University of Chicago

PREFACE

The four lectures printed here were presented at the University of Coimbra on the 27th and 29th of April and the 2nd and 4th of May, 1983. A fifth lecture, based partially on the first two was given at the Academia das Ciências in Lisbon on may 5th.[1] That these lectures were given at all is due in large measure to Professor A. J. A. de Gouveia, former Rector and former Chairman of the Department of Chemistry at the University, whose interest in the history of chemistry led first to a correspondence on this subject and then to a joint invitation from the Department of Chemistry and the University as a whole.

That these lectures are historiographical in orientation is due to a long standing interest of mine in this subject.[2] Any historian of science or medicine educated in the 1950s is well aware of the many changes that have taken place in this subject since that time. The positivism of George Sarton was under attack even before his death in 1956 and today there are many competing approaches, methodologies and interpretations from which scholars may choose. The effects of reading such differing accounts on our students led me to introduce a required course on historiography for our first year graduate students in the history of science at the University of Chicago fifteen years ago. It was my hope to expose these students to the broad spectrum of writing available in the history of science and medicine so that they might be better able to orient themselves in the field. I hoped also to teach them to become more tolerant of the views of others by showing them how frequently historical interpretations have changed in the past.

I do, in fact, feel that the greatest need in the history of science and the history of medicine at the present time is tolerance. At present we see increasing inflexibility as individuals move into new areas of research that would not have been considered significant one or more generations ago. As these scholars uncover important material they tend to become impatient with those still wedded to more traditional views. But those who teach the history of science and medicine must realize that they do not exist to turn out carbon copies of themselves. Rather, it is their job to take potential scholars from any relevant field and to try to give them additional tools they need so that they may make some original contribution. If they have been trained as scientists and wish to continue in the accepted internalist traditions that is to be encouraged. If they come from some other area relating science to another aspect of our culture or society they too should be welcomed. Indeed, the traditional approach does present us with the necessary framework of our discipline. However, the newer interpretations will eventually make this field acceptable to historians who can see the importance of science and medicine for world history, but who frequently despair of ever understanding the technical papers they meet so often in the literature of the history of science.

It is then my belief that a study of the history of the history of science and medicine is important for the training of students who hope to become professionals in this field. But these lectures also call attention to the historical significance of one particular science, chemistry, and to my own opinion of the role of chemical history in the development of the Scientific Revolution. My invitation to Coimbra was closely connected with the fact that a course on the history of science has only recently been instituted there through the efforts of Professors A. J. A. de Gouveia, A. M. Amorim da Costa, and the Department of Chemistry. Their interest was centered primarily on the history of chemistry. For this reason my lectures draw heavily on my own background and my research.

Trained as a chemist, I was drawn to the history of science because I wanted to learn more of the role of chemistry in the period of the Scientific Revolution. However, when I became a student in the history of science at Harvard University I soon found that chemistry was generally dismissed as a relatively insignificant factor in the development of modern science. I was then faced with an historiographical problem of my own... to demonstrate that this relatively neglected topic did deserve attention by scholars. For this reason my research over the years has been aimed at modifying the internalist tradition of Alexandre Koyré which dominated the period of my graduate study.

It was then for a variety of reasons that for the subject of these lectures I decided to discuss first the development of the history of science as a field and then turn more specifically to the history of chemistry and its importance for our understanding of the Scientific Revolution. Accordingly, in the first lecture I rapidly traced the writing of the history of science and medicine from the Renaissance to the work of George Sarton indicating the extent to which history may reflect the deeply rooted beliefs of historians. The second lecture discussed the period since the death of Sarton pointing out new areas of research and challenges to the positivistic tradition. In my third lecture I turned specifically to the history of chemistry to show how its development differed from that of the history of science in general. And in the final lecture I have drawn primarily from my own earlier research to show that the development of chemistry in the sixteenth and seventeenth centuries is indeed an essential part of the Scientific Revolution.

These comments will already have led the reader to the conclusion that these lectures present a very personal point of view. This is certainly true. I do not think it could be otherwise. I had no intention of making them comprehensive in nature and I am sure that many would prefer to have seen the works of other authors discussed than those I have chosen. I feel certain also that many readers would paint a very different picture of the development of this field than I have. I would expect that, and, indeed, hope that it is so because this is one of the lessons to be drawn from the literature of this field. After all, we do reflect our own backgrounds and in a field as diverse as the history of science there are no two individuals with the same scientific, philosophical and historical training. — Let me add one final *Caveat:* the reader will find an emphasis on English and American authors. I make no apology for this. The impact of Sarton above all affected the growth of the history of science in the United States. It was here and in England where I was trained and have carried out most of my research. These authors, moreover, are the ones who are most accessible to our students.

Finally let me thank Professors de Gouveia and da Costa and their families for their many kindnesses to me and my wife while we were in Portugal. I also wish to extend my thanks to the Rector of the University, Professor Rui Alarcão, the Vice-Rector, Professor Jorge S. Veiga, the Chairman of the Department of Chemistry, Professor José Simões Redinha, and to all the members of that Department and the others who attended these lectures. I am also indebted to the President of the Academy of Science of Lisbon, Professor José Pinto Peixoto, for his friendly welcome and to the many members of the Academy we met while we were there. Indeed, it is because of the people we met no less than the beauty of the country that our visit to Portugal will always remain a most memorable experience.

In a larger sense I am indebted to the many lively discussions I have had with the students in my course on the historiography as they struggled with the texts described here over the past fifteen years. I must also thank Professor Joe D. Burchfield of Northern Illinois University for reading and commenting on the first three lectures in their first draft... and to John Neu at the University of Wisconsin-Madison for making available to me several rare volumes which were not available at the University of Chicago.

In 1976 Dr. I. Grattan-Guinness — then the Editor of *Annals of Science* — invited me to prepare an article for a series he planned for this journal on «Personal Viewpoints on the History of Science.» Other commitments made it impossible for me to complete my contribution to this series, but I did make some notes for the paper and should he ever chance to read these pages, he may be pleased to know that these have been incorporated into this preface and the lectures themselves. [3]

Finally I wish to mention Walter Pagel. His friendship over a period of nearly a quarter century led to hundreds of discussions at his home in Mill Hill which profoundly influenced my own views on this subject. Oddly, it was while I was in the midst of writing my own assessment of the impact of his work for these lectures that I received a telephone call from Professor P. M. Rattansi of University College, London, informing me of Professor Pagel's death on March 25th, 1983. I gratefully acknowledge my debt to him and dedicate these lectures to his memory.

Allen G. Debus
Deerfield, Illinois, U.S.A.
14 September 1983.

NOTES

1. To be published as «The History of Science Today» both in the *Memórias da Academia das Ciências de Lisboa* and in a separate volume on the history of science in Portugal to be published by the Academy.

2. Over the years this interest has led to a number of papers related to the history of the history of science. Among them may be cited «An Elizabethan History of Medical Chemistry,» *Annals of Science 18* (1962, published 1964), 1-29; «The Significance of Early Chemistry,» *Cahiers d'Histoire Mondiale 9* (1965), 39-58; «Alchemy and the Historian of Science,» *History of Science 6* (1967), 128-38 [Essay review of C. H. Josten's *Elias Ashmole];* «The Chemical Philosophy of the Renaissance» in *The Rise of Modern Science: Internal or External Factors?,* George Basalla (ed.) (Lexington, Mass: D. C. Heath, 1968), pp. 82-88 [this volume is a useful collection of selections from Whewell to the present relating to differing interpretations of the Scientific Revolution. It should be reprinted]; «Chemistry and Scientific Revolution» in *Teaching the History of Chemistry. A Symposium, San Francisco, California U.S.A., April, 1968* (Budapest: Akadémiai Kiadó, 1971), pp. 101-11; «The History of Chemistry and the History of Science,» *Ambix 18* (1971), 169-77; «The Relationship of Science-History and the History of Science,» *The Journal of Chemical Education 48* (1971), 804-5; «Some Comments on the Contemporary Helmotian Renaissance,» *Ambix 19* (1972), 145-50 [Essay review of the reprint of J. B. van Helmont's *Aufgang der ArtzneyKunst* (1683; 1971)]; «The Pseudo-Sciences and the History of Science,» *The University of Chicago Library Society Bulletin 3* (1978), 3-20; «The Arabic Tradition in the Medical Chemistry of the Scientific Revolution» in the *Proceedings of the First Internacional Symposium for the History of Arabic Science, April 5-12, 1976,* Ahmad Y. Al-Hassan, Ghada Karmi and Nizar Namnun (eds.) (Aleppo: Institute for the History of Arabic Science, 1978), *2,* pp. 275-90; «The Geber Tradition in Western Alchemy and Chemistry» in the *Proceedings of the Second International Symposium for the History of Arabic Science, 5-12 April 1979* (Aleppo: The Journal of the History of Arabic Science, in press); Obituary of Walter Pagel, *Bulletin of the History of Medicine* (in press).

3. Cyril Stanley Smith's contribution to Grattan-Guinness' series was titled «A Highly Personal View of Science and its History» [*Annals of Science 34* (1977), 49-56], a title that would well suit my own series of lectures for the University of Coimbra.

LECTURE 1

SCIENCE AND HISTORY: THE BIRTH OF A NEW FIELD

When I was an undergraduate student at Northwestern University thirty-five years ago the history of science did not exist as a field of study. My major field was chemistry, and while it is true that a short course of lectures on the history was offered through that Department, few students knew of it and fewer still attended these lectures. At the time I had a minor field in history, but as far as the members of that Department was concerned, the subject was one to be avoided. To be sure, it was acknowledged that there had been a Scientific Revolution, but what its meaning might have been for world history was never told to us. Nor was reference made to the science or technology of the past hundred years. The standard teaching fare continued to emphasize politics, society and religion. Even the few lectures we heard on intellectual history seemed to ignore the sciences. The student of history was left with the impression that world history had been untouched by the rise of the sciences and medicine while for the student majoring in any of the sciences the subject was to be ignored for fear of cluttering the mind with obsolete theories and facts.

Of course I see now that my impressions as a student were wrong. The study of the history of science and medicine is important. Furthermore, although I did not know it then, I know now that the writing of the history of science and medicine has a long history. This is true even for the ancient period. The short histories of geometry and medicine written by Proclus and Celsus are well known.[1] Nor did this tradition expire in the Middle Ages. Gui de Chauliac prefaced his famous *Surgery* (1343) with an historical survey of the field.[2] True, there are not many of these early histories, but there are enough of them to indicate that it was considered important to know not only the subject matter of one's field, but its historical development as well.

It would be a matter of considerable interest to discuss the writing of the history of science and medicine in detail, but there is hardly enough time to do this in the short time that I have. Today I will range over nearly four hundred years, from the late sixteenth through the mid-twentieth centuries, but I will do so only to make a few points. I would like to show that there is a difference in the types of scientific histories written before and after the great scientific watershed of the midseventeenth century. Indeed, the scientific beliefs of the authors affected their historical outlook. I would like to touch further on the scientific histories of the Enlightenment and the early nineteenth centuries to indicate their influence which extended into the present century.

Scientific History in the Renaissance: The Paracelsians and Antiquity

Let me turn first to the Scientific Revolution. It was during the sixteenth and seventeenth centuries that we see for the first time a growing number of historical writings in the sciences. These show us one of the most characteristic features of historical writing in all periods, that the historian writes with a purpose in mind. Indeed, the historian is often a propagandist even when he may not be aware of it himself. In the late 1960s we witnessed the birth of a new school of radical historians whose work fit the temper of the times. In the sixteenth and seventeenth centuries we see a similar situation. The move for reform in religion and science was then reflected in the writings of contemporary historians. Thus, the medical and chemical reforms of Paracelsus (1493-1541) were to be seen in the works of his followers.[3] These reforms were both theoretical and practical in nature. On the one hand they sought a new understanding of the world based on a mystical view of the cosmos which they interpreted through man the microcosm and the all encompassing macrocosm. Everything in the small world of man was to be found in the great world of the macrocosm. And man, as a true natural magician, was to learn of his Creator through the study of God's created nature. Chemistry was to be the key to this new knowledge since both nature and man were best understood through chemical processes and analogies. Therefore, if the Paracelsians argued for a new theoretical basis of knowledge, they also saw a need for practical reform because human physiology was described in chemical or alchemical terminology. New chemically prepared medicines were used to combat what were thought to be chemical disorders in the body. These were considered to be far more useful than the traditional Galenical herbal mixtures. In short, the Paracelsian medicine and science of the sixteenth century was anti-Galenic and anti-Aristotelian. It called for fresh observations and it was argued that these could be interpreted properly only by chemists for the benefit of physicians.

This message was elaborated on by a number of authors who used history to support their convictions. Let me illustrate this through the works of two sixteenth-century authors. The first of these was interested primarily in the theoretical aspects of the Paracelsian Philosophy while the second confined himself primarily to the practical aspects of the introduction of the new chemically prepared medicines. The title of the first work is long and typically descriptive for the period: *The difference betwene the auncient Phisicke, first taught by the godly forefathers, consisting in vnitie peace and concord: and the latter Phisicke proceeding from Idolaters, Ethnickes, and Heathen: as Gallen, and such other consisting in duality, discorde, and contrarietie* (1585). The author, R. Bostocke, Esquire, attacked the false natural philosophy and medicine being taught at the universities. Although the new chemical medicine could be proven to be valid by experience, how were students to know this?

> ...in the scholes nothing may be receiued nor allowed that sauoreth not of *Aristotle, Gallen, Auicen,* and other Ethnickes, whereby the yong beginners are either nor acquainted with this doctrine, or els it is brought into hatred with them. And abrode likewise the Galenists be so armed and defenced by the protection, priuiledges and authoritie of Princes, that nothing can be allowed that they disalowe and nothing may be receiued that agreeth not with their pleasures and doctrine... [4]

With Bostocke as with other writers of this period religion was a major factor. Aristotle and Galen were heathens. Their false philosophy and medicine had been perpetuated by lecturers who read and commented on their texts without any search for confirmation. Rather than the books of the ancients, the seeker of truth should learn from God. «The Almighty Creatour of the Heauens and the Earth (Christian Reader), hath set before our eyes two most principall Bookes: the one of Nature, the other of his written Word...» [5]

Here was a different approach to nature calling for the destruction of the philosophy of the ancients and its replacement by a new science based upon Holy Scripture, observation and experiment. This would be a Christian philosophy.

For Bostocke history was an important tool and he devoted nearly one half of his treatise to the history of chemistry and medicine.[6] Believing that the pristine knowledge granted to Adam could be recovered from the Old Testament and the Corpus Hermeticum which was thought to be almost as old, he argued that these truths had been partially preserved in the writings of the pre-Socratics and Plato. But Aristotle had attacked his teacher while Galen who had adopted the philosophy of Aristotle had compounded his sins by persecuting Christians. In the succeeding centuries only a dedicated few — for the most part Byzantine and Islamic alchemists, had preserved the most ancient truths which they passed on from master to pupil. Thus, Paracelsus was not an innovator. Rather, his reforms in medicine were properly to be compared with the reforms of Copernicus who had rediscovered the true ancient astronomy and the reforms of Luther, Melanchthon, Zwingli and Calvin who had rediscovered the theological truths of antiquity. [7]

It was to be expected that such accounts were to be attacked by those who considered the work of Aristotle and Galen the glory of the learned world. Such an author was Thomas Erastus (1524-1583) who pictured Paracelsus as an ignorant charlatan who preferred magic and the devil to the classical authorities. More interesting, however, is the reaction of those who sought compromise. Johann Albertus Wimpenaeus (1569) saw value in the work of Paracelsus, but he did not reject the ancients since wisdom was to be found in both. [8] Guinter of Andernach (c. 1505-1574) is even more interesting as a man who was perhaps the most famous of the Renaissance medical humanists. [9] As a young scholar he prepared translations of much of Galen as well as Paul of Aegina, Oribasius and Alexander of Tralles. As Professor of Medicine in Paris he taught Andreas Vesalius and Michael Servetus, both of whom served as his assistants.

In many respects Guinter represents the quintessential scholar for he never ceased to study or to learn. Thus, we find that as he grew older he carefully read the new Paracelsian medical works. But how were these to be assessed by a learned Galenist? Guinter's answer is to be found in his massive *De medicina veteri et nova* published in 1571. Here it may be seen that he had formed a high opinion of the chemical medicines that were to be essential to Paracelsian authors. Indeed, he wrote that the «medicines of the chemists are more than divine.» [10] Still, this humanist insisted that the theoretical basis of medicine must remain grounded on Galenism. Paracelsian thought was offensively mystical and its proponents were arrogant.

Guinter's problem was to dismiss Paracelsus and his fanatical disciples while retaining the benefits of the chemical medicines. This goal was quite different from the goal of a Paracelsian such as Bostocke, but like Bostocke, Guinter turned to history for his answer. [11] The earliest men, he wrote, were strong and needed only simple and mild remedies for retaining their health. More debilitating diseases arose only later when the accumulated luxuries of centuries resulted in a permanent corruption of mankind. It is then that we see different medicines — resins and aromatic substances — introduced in the texts of the Islamic and Indian authors. It has been the destiny of Paracelsus not only to restore to use chemicals known to other authors, but also to enrich them with a treasury of new waters, liquors, salts and oils — medicines often more efficacious than the traditional ones. Guinter's answer to the current medical debate therefore was compromise. Both medicines were needed. «The ancients on account of time honored authority are to be given first place,» but there was much of great value in the work of the more recent chemists. Would that Galen had been more brief and more accurate; would that Theophrastus [Paracelsus] had been more open and candid! There are faults and virtues in the work of both factions, and physicians must choose the best from each. [12]

In contrast with Bostocke, Guinter of Andernach did not seek to recover a pristine medicine known to Adam as a means of proving the antiquity of Paracelsian truths. Instead, the ancient art of medicine was once more ascribed to Greek authors. Chemical medicines had been introduced by the Arabic physicians. They were later forgotten and eventually rediscovered by Paracelsus. This had been a notable achievement and he was to be praised for it. But his mystical cosmology could be safely ignored — or else simply shown to be a restatement of concepts known to the ancient Greeks.

Scientific History in the Enlightenment

Had the course of the Scientific Revolution followed the path charted for it by the Paracelsians, the Hermetic history of Bostocke might still be read today. But it did not. The triumph of the mechanists in the seventeenth century placed an emphasis not on chemistry and medicine, but on astronomy and the physics of motion. This change may also be followed in contemporary histories. Mechanists of the late seventeenth and the eighteenth centuries sought to divorce themselves from the mysticism and the magic of their predecessors no less than their dependence on the Greek philosophers. But when they wrote of the warfare between the «ancients» and the «moderns» they specifically meant the adherents of the Greek philosophers versus the Mechanical philosophers. These mechanists found atomism a useful tool as they sought an explanatory model based upon the size, shape and motion of the small parts of matter. Mathematical abstraction became in their hands a powerful tool for the analysis of natural phenomena and the *Principia mathematica* (1687) of Isaac Newton was almost to become the Bible of the new science.[13]

Because of the importance of Newton for the eighteenth century it is of some interest to pause for a moment to look at the work of John Freind (1675-1728). Freind was a disciple of Newton who held the chairs of chemistry and medicine at Oxford. He published both on chemistry and medicine, and his views, as we should expect, were colored by the science he believed in. His chemical work was an open attempt to divorce himself from earlier chemical authors. Indeed, he sought to explain chemical reactions as the interaction of spherical atoms possessed of forces similar to the gravitational forces Newton had postulated for the solar system. This was an attempt to establish a Newtonian chemistry.[14]

But it is in Freind's history of medicine published in 1725 and 1726 that we can compare his views with those of Bostocke and the earlier Paracelsian apologists. For Freind the mystical religious outlook of the Paracelsians could not be tolerated. Freind rejected Paracelsus as an idle systematizer whose whole cosmology and religious-vitalistic outlook toward nature were the very antithesis of the new science. On the other hand, by the opening years of the eighteenth century there seemed to be little doubt of the value of chemically prepared medicine. However, Freind would not permit Paracelsus to be credited with their discovery. Rather, like Guinter, he insisted that the honor of their discovery be given to Arabic chemists and physicians.[15] ....How very different are the histories of Freind and Bostocke! They are both histories of medicine, but written from two opposing viewpoints separated by the scientific watershed of the midseventeenth century.

For the philosophers of the eighteenth century the example of Newton and the new science signified the birth of a new age. The historian was told to abandon his traditional studies. What moral value was to be found in the story of kings, popes and wars? Far different was the history of science since in it we witness the ennobling example of human progress wrought from ignorance by the true heroes of mankind. Late in the century a journalist wrote that

> The incredible discoveries that have multiplied during the last ten years... the phenomena of electricity fathomed, the elements transformed, the airs decomposed and understood, the rays of the sun condensed, air traversed by human audacity, a thousand other phenomena have prodigiously extended the sphere of our knowledge. Who knows how far we can go? What mortal would dare set limits to the human mind? [16]

Not only would science continue to progress, perhaps its history would permit us to forecast the future. As Gelbart has written,

> Toward the end of the century this faith in the internal dynamism of science becomes much more explicit and encourages a very optimistic picture of the future. The same force that propelled man to his present state of scientific sophistication and technical prowess still operates, and will spur him on in the future. While progress may not always be steady and even, it will eventually take man to intellectual heights which he cannot even imagine. The scientific past, then, offers valuable clues for the future. The search for truth is not ended. Great as we are, our science and the science of our children will be ever surpassed. [17]

Belief in being able to plan the future was to prove an unfulfilled hope, but a deep seated belief in the importance of scientific history as evidence of progress was to become a characteristic of eighteenth- and nineteenth-century science. Joseph Priestley's histories of electricity and the gases of the air, J. E. Montucla's history of mathematics and Jean Sylvain Bailly's various histories of astronomy are serious studies still referred to by scholars.[18] This is a period when scientists began to add histories of their subjects to scientific treatises as we see in Boerhaave's *New Method of Chemistry* and Laplace's *System of the World.* [19]

And yet, in its own way, this was as biased a history as that of the Renaissance Paracelsians. These historians wrote of progress in the past leading to the current state of the sciences. The emphasis was invariably placed on the science of Western Europe. The religious aura of the period of the Middle Age was treated with contempt and blamed for the lack of progress in that period. Little more consideration was given to the accomplishments of the Far East or Islam.

Science and Religion in the Nineteenth Century

The Enlightenment view that science is essencially progressive has left its stamp on the field to our own day and it is only in recent decades that there has been a search for a broader historical context in which to place scientific history. Again let me recall my own graduate training. Then we were told that the first true history of science — that is, a history of science as a whole in contrast to the histories of individual sciences so common in the eighteenth century — was to be found in William Whewell's *History of the Inductive Sciences* (1837). Whewell was one of the great physicists of Victorian England and his title reflects his belief in the ideal of Baconian science, a science that is primarily inductive rather than deductive — one that is based upon observation and experiment. Like Bacon, Whewell sought to avoid a science that was based too heavily on mathematics. Also like Bacon, he felt that the primary goal of the history of science was to furnish the subject matter for the philosophy of science. History for Whewell was subordinate to philosophy, or to put it another way, our goal as historians is to elucidate scientific method.[20]

Whewell's history leaves great areas untouched. He ignored the achievements of the ancient Near East partially because solid information was sketchy when he wrote, but also because he felt that the science of Egypt and Babylonia was devoid of theory. The accomplishments of the Far East and Islam fared little better and he even wrote harshly of Greek science because of its deductive nature. As for his description of the science during and after the Scientific Revolution it is easy to recognize his dependence on the viewpoint of the Enlightenment. Separate chapters are devoted to each of the sciences or current areas of research. It was certainly not an integrated history and, because of Whewell's personal interests, there is little to be found here on the biological or medical sciences.

It is in Whewell's chapters on the Middle Ages that his prejudices are most evident.[21] For him this was «the Stationary period» resulting from Christianity which caused a neglect of physical reasoning. The so-called scientists of the period had added nothing new to knowledge. He derided them for their «indistinctness,» «dogmatism,» «mysticism,» and their «commentatorial spirit.» To be sure, he made a few exceptions — Roger Bacon and the architects of the cathedrals come to mind, but in general he dismissed a millennium of history with distaste and the belief that during that time the physical sciences had become little more than magic. With evident relief he turned his back on this period of dogmatism.

> The causes which produced the inertness and blindness of the stationary period of human knowledge, began at last to yield to the influence of the principles which tended to progression. The indistinctness of thought, which was the original feature in the decline of sound knowledge, was in a measure remedied by the steady cultivation of Pure Mathematics and

> Astronomy, and by the progress of inventions in the Arts, which call out and fix the distinctness of our conceptions of the relations of natural phenomena. As men's minds become clear, they become less servile: the perception of the nature of truth drew men away from controversies about mere opinion; when they saw distinctly the relations of *things,* they ceased to give their whole attention to what had been *said* concerning them; and thus, as science rose into view, the spirit of commentation lost its way.[22]

Wherever we look among nineteenth-century historians we find views similar to those of Whewell. W. E. H. Lecky wrote of the *History of the Rise and Influence of the Spirit of Rationalism in Europe* (1865), Georges Cuvier titled a survey of recent work at the Parisian Academy of Sciences the *Histoire des progrès des sciences naturelles depuis 1789,* and of course, Auguste Comte's positivism was based upon scientific progress. All those imbued with this essentially Enlightenment attitude were faced with the contrast between medieval and early modern science. As Whewell had noted, the first coincided with a period of the dominance of the Roman Catholic Church. Others were to make much of the fact that the Scientific Revolution occurred during the Reformation and its aftermath. Both cases forced scholars to look at the relationship of science to religion and this was to become a subject of debate beginning in the late nineteenth century. This may be well illustrated in the works of many authors, but it is seen with great clarity in the works of three, John William Draper, Andrew D. White, and James Walsh.

John William Draper (1811-1882) was a prominent American chemist and physiologist.[23] He was well known as a student of photography and a participant in the famous Oxford debate of 1860 that pitted T. H. Huxley against Bishop Wilberforce over the truth of Darwinism. Draper was a disciple of Auguste Comte and devoted much of his effort in later years to the study of history. His *A History of the Intellectual Development of Europe* (1863) is an important example of nineteenth-century intellectual history while his *History of the Conflict Between Religions and Science* (1874) has become one of the most widely read books of the past century. By 1910 it had reached a twenty-fifth edition and it has been reprinted as recently as 1972. There is probably no other work in the history of science that can match this record.

In both of these works Draper upheld the validity of science over that of religion, but the more moderate tone of the first text was lost in his *History of the Conflict Between Religion and Science.* The reason? ...Draper's reaction to the Vatican Council of 1869-1870 which declared the supremacy of the Pope and included canons such as the following:

> Let him be anathema —
>
> Who shall say that human sciences ought to be pursued in such a spirit of freedom that one may be allowed to hold as true their assertions, even when opposed to revealed doctrine.
>
> Who shall say that it may come to pass, in the progress of science,

> that the doctrines set forth by the Church must be taken in another sense than that in which the Church has ever received them and yet receives them. [24]

For Draper the aspirations of the Papacy threatened to renew the Dark Ages. He believed that the demand for unquestioned belief in things above reason had ended the scientific advance of antiquity. A scientist required something far different — a belief that the universe is governed by immutable law. The mind of the scientist must be open to all possibilities in his quest for truth, not tied to blind faith!

Andrew Dickson White (1832-1918), was a prominent American diplomat and educator who served as U. S. minister to Germany and Russia as well as Head of the U. S. Delegation to the Hague Peace Conference (1899). More important for our story is the fact that White organized Cornell University with Ezra Cornell and then went on to serve as its first President. In contrast to most universities at that time White sought a non-sectarian institution that would serve as a refuge for the sciences and humanities. He was both shocked and surprised to find strong opposition to this goal from members of organized religions. White hoped first to convince his adversaries through reason, but eventually he delivered a lecture (1875) on «The Battlefields of Science» in which he took as his thesis that

> In all modern history, interference with science in the supposed interest of religion, no matter how conscientious such interference may have been, has resulted in the direst evils both to religion and to science, and invariably; and, on the other hand, all untrammelled scientific investigation, no matter how dangerous to religion some of its stages may have seemed for the time to be, has invariably resulted in the highest good both of religion and of science. [25]

Reaction to his lecture was immediate and encouraging. He was soon invited to speak on the subject at numerous university associations and literary clubs. The lecture was expanded and published as a small book, *The Warfare of Science,* and he continued his work on the subject to contribute «New Chapters in the Warfare on Science» to *The Popular Science Monthly.* In the meantime he had seen Draper's *Conflict Between Science and Religion.* His first thought had been that nothing more need be added to the subject until he realized that Draper «regarded the struggle as one between Science and Religion. I believed then, and am convinced now, that it was a struggle between Science and Dogmatic Theology.» [26]

In many ways White's history (published in final form in 1895) is similar to that of Draper's. He deplored the impact of *Genesis* on the history of science, he argued that a belief in the imminent end of the world is useless for the growth of science, and inveighed against a dogmatic reading of Scripture. Draper had aimed his work primarily at the Roman Catholic Church. White agreed, but his experience at Cornell had taught him that the Protestant Churches were no more liberal in these matters. Above all, he returned again and again to the Enlightenment which he saw as a period of the conflict of reason and mystery. And if the sciences have managed to progress, he believed that the complete triumph of reason over mystery had not yet occurred.

The case for Roman Catholicism was to be made by James Joseph Walsh (1865-1942), professor at Fordham Medical School and later the Director of the Fordham School of Sociology whose *The Thirteenth, Greatest of Centuries* (1907; 14 ed., 1952) and *The Popes and Science* (1908) were directed against White and Draper. Walsh wrote of the importance of the Papacy as a patron of the sciences and education in the medieval period. As a physician Walsh emphasized the important anatomical work carried out in the Italian universities in the late Middle Ages and characterized this work as essential background for the work of Vesalius and Harvey. He also pointed to specific work in chemistry and physics while he attributed the experimental method to thirteenth century authors, primarily Roger Bacon and Albertus Magnus. It was the Church that had established hospitals and universities while in contrast, the Reformation

> had carried away with it in its course nearly everything precious that men had gained during the four centuries immediately preceding. Art, education, science, liberty, democracy — everything worth while had been ruined for the time. [27]

As for the Enlightenment of the eighteenth century,

> The fact of the matter is that... there was a great decadence of interest in scholarship and true education. There is a distinct descent in human culture at this time. Education was at its lowest ebb, hospitals were the worst ever built, art and architecture were neglected, and human liberty was so shackled that the French Revolution was needed to lift the fetters from men's minds as well as bodies. [28]

Walsh had written originally to refute the views of Andrew Dickson White. As for Draper he dismissed his work as little more than a «comic history» founded on ignorance. [29]

The Draper-White-Walsh debate was more than a dispute over science and religion. More important was the fact that it called for a new assessment of science and medicine in the Middle Ages. Here Walsh was to be vindicated. Nineteenth-century editions of medieval medical texts led the way toward a new appreciation and by the early years of the new century there were a host of new works — many of them still useful — penned by Max Neuberger, Julius Pagel, Karl Sudhoff and others. [30] Indeed, medical history matured in

Germany and remained its center until well after the first World War. At the same time, in the physical sciences, Moritz Cantor was at work on his four volumes *Voreslungen über Geschichte der Mathematik* (1880-1908), while Paul Tannery, Sir Thomas Little Heath and Johan Ludvig Heiberg prepared their monumental editions of Greek mathematicians and early modern scientists. The great French physicist and philosopher of science, Pierre Duhem, carried out his research which led to his ten volume *Le Système du Monde* (10 vols.; 1st five vols., 1913-1917) a work which led to a completely altered view of medieval science and a debate over the originality of Galileo that is not yet settled.

George Sarton: The Establishment of a New Academic Field

In short, by the early years of the twentieth century there was a conscious realization among a small number of scholars that scientific and medical history was not only interesting, it was essential for understanding history as a whole. This was clearly a period of ferment and it was the time that George Sarton (1884-1956) was a student. A Belgian, he received his B. Sc. in 1906 and his Sc. D. in 1911, both from the University of Ghent. However, most of his carreer was spent in the United States because he left his homeland during the first World War.

Sarton was trained as a mathematician, but his interests embraced all of the sciences. He was a dedicated man who believed that his was the most valuable form of history. To this end he took part in the founding of societies for the history of science throughout the world. He also worked for the establishment of scholarly journals in the field. He was the founding father of the History of Science Society and *Isis* — still the best known journal in the field — which was first published in 1912. He wrote an enormous number of books, articles and reviews. Perhaps the best known of these is his *Introduction to the History of Science* which began with Homer and eventually reached the fourteenth century in three massive volumes in five parts. This work was published over a twenty year span and finally given up only when Sarton realized that it would be impossible to continue because of the vast amount of material still to cover. He also planned the publication of the lectures that composed his two year survey of the field that he gave at Harvard University — only two of the planned eight volumes had been completed by the date of his death. In the short time I have here it would be impossible to list even his major writings. Nor is it necessary to do so for our purposes. However, it is necessary to say a few words about his approach to the field because of his enormous influence.

George Sarton frequently expressed his debt to the writings of Auguste Comte and there is no doubt that he considered himself a positivist. Writing in 1927 he defined science as «systematized positive knowledge.» [31]

> Our main object is not simply to record isolated discoveries, but rather to explain the progress of scientific thought, the gradual development of human consciousness, that deliberate tendency to understand and to increase our part in the cosmic evolution.[32]

In the *Introduction* he had little to say about ancient science prior to the Greeks because he felt that Oriental science was largely devoid of theory. On the other hand he felt that his work contained the first real account of medieval science. This is doubtful because as late as 1927 he made no reference to Duhem's *Système du monde* which was to transform our views of medieval physics in a way that Sarton's never was to do. By the time of the publication of the first volume of his Harvard lectures (1952) he corrected both omissions.

Sarton did not retreat on other points. As a positivist he wished a history of real science — that is, science as we know it today. Subjects outside of science that may have formed part of man's outlook to nature in earlier periods were ignored or branded as «pseudo-sciences.» We know now that alchemy and natural magic were important elements in the development of modern science. Sarton was willing to accept the actual chemical reactions and equipment described by the alchemists in his history of science, but nothing else.

> The historian of science can not devote much attention to the study of superstition and magic, that is, of unreason, because this does not help him very much to understand human progress. Magic is essentially unprogressive and conservative; science is essentially progressive; the former goes backward; the latter, forward. We can not possibly deal with both movements at once except to indicate their constant strife, and even that is not very instructive, because that strife has hardly varied throughout the ages. Human folly being at once unprogressive, unchangeable, and unlimited, its study is a hopeless undertaking. There can not be much incentive to encompass that which is indefinite and to investigate the history of something which did not develop.[33]

A second point is that Sarton maintained the division that had developed between the history of science and the history of medicine. As we have seen, the history of medicine had developed independently in the late nineteenth and the early twentieth centuries. Henry Sigerist, a pupil of Sudhoff, left Leipzig late in the 1920s to take over the directorship of the recently founded Institute for the History of Medicine at The Johns Hopkins University in Baltimore. There he hoped to continue the German tradition in the United States. But for Sarton the claims of the historians of medicine posed a threat to his fledgling history of science. He believed that there was a hierarchy of the sciences. Mathematics stood at the top since it was necessary for the mathematical sciences: astronomy, physics, and chemistry. Only eventually as we followed this scheme would we descend to the life sciences. He explained that

> Men understand the world in different ways... some men are more abstract--minded, and they naturally think first of unity and of God, of wholeness, of infinity and other such concepts, while the minds of other men are concrete and they cogitate about health and disease, profit and loss. They invent gadgets and remedies; they are less interested in knowing anything than in applying whatever knowledge they may already have to practical problems; they try to make things work and pay, to heal and teach. The first are called dreamers...; the second kind are recognized as practical and useful. History has often proved the hortsightedness of the practical men and vindicated the «lazy» dreamers; it has also proved that the dreamers are often mistaken.
>
> The historian of science... is not willing to subordinate principles to applications, nor to sacrifice the so-called dreamers to the engineers, the teachers, or the healers. [34]

Sarton surely idolized the dreamers. And, as he believed that the biological sciences stood far below the mathematical sciences, he believed that medicine was lower still. Because he was convinced that medicine was a practical art he was distressed by those medical historians who claimed that medicine is the real foundation of the other sciences. Indeed, he wrote, «the main misunderstandings concerning the history of science are due to historians of medicine who have the notion that medicine is the center of science.» [35] Sarton felt that medical historians had presented a warped version of scientific history because of their insufficient scientific knowledge. We need not be too surprised that societies of medical history seldom meet jointly with societies of the history of science even today... or that for decades the history of science centered on the physical rather than the biological sciences.

I have covered four hundred years of the writing of the history of science and medicine in abstracted form hoping to make a few general points. The first of these is almost obvious, that historians write with a purpose — and it may frequently become propaganda for their own deeply held beliefs. Thus we looked briefly at a few histories written in the sixteenth century. For the followers of Paracelsus scientific and medical truths were closely allied with their religious convictions, their desire to overturn the educational establishment and their belief that the truths of the Paracelsian medicine could be connected with divine truths imparted to Adam at the Creation. Their histories were written to establish their medicine and natural philosophy.

The success of the mechanical philosophy in the course of the seventeenth century led to a different historical model, one in which religion was divorced from scientific progress. Enlightenment philosophers, looking on the science of the Middle Ages as a barren period of science erased a millennium of science from their works pausing only to refer to Roger Bacon and a few other authors who seemed to rise above the general intellectual malaise. Indeed, during this period, the history of science was consciously divorced from the traditional historical subjects since science represented human progress while politics, warfare and religion did not. This essentially Enlightenment approach to the sciences dominated the history of science well into this century. Walsh's reply to the histories of Draper and White stands as one of the early defenses of religion and medieval science and medicine. Indeed, it is not until we reach the closing decades of the nineteenth century that we begin to see an important series of texts and historical analyses that were to reshape our views of medieval science.

Although there are many works on the history of science and the history of medicine that were published prior to the first World War, the field was not yet established academically and it is for this reason that George Sarton is such an important figure. This is surely due more to his efforts than those of any other individual. He not only established the field at Harvard University, he founded the journal *Isis* and is the person most responsible for organizing the international History of Science Society. He diligently sought out others with similar interests and encouraged them to pursue their work. It is little wonder that with Sarton's world-wide activities that by the middle decades of this century the field reflected his prejudices. In the post-World War II era a few Ph.D.s were being graduated from the new field at Harvard University and elsewhere, but the work of these young scholars as well as the pages of *Isis* still reflected a positivistic approach emphasizing the development of the physical sciences. This was to change radically in the following years.

NOTES

1. Morris R. Cohen and I. E. Drabkin, eds., *A Source Book in Greek Science (Cambridge: Harvard U.P., 1958), pp. 33-38 (Proclus), 468-73 (Celsus).*

2. *Gui de Chauliac's history of surgery is conveniently available in The Portable Medieval Reader,* James Bruce Ross and Mary Martin McLaughlin, eds., (New York: The Viking Press, 1973), pp. 640-49.

3. My views on the Paracelsian tradition are elaborated in *The Chemical Philosophy: Paracelsian Science and Medicine in the Sixteenth and Seventeenth Centuries, 2 vols., (New York: Science History Publications, 1977).*

4. *R. Bostocke, Esq., The difference betwene the auncient Physicke... and the latter Phisicke* (London: Robert Walley, 1585), sig. Fii^{v}.

5. Thomas Tymme, *A Dialogue Philosophicall* [London: T. S. (nodham) for C. Knight, 1612), sig. A3.

6. Bostocke's history is reprinted with an introduction and annotations by Allen G. Debus in «An Elizabethan History of Medical Chemistry,» *Annals of Science 18* (1962, published 1964), 1-29.

7. See Debus, *The Chemical Philosophy 1,* pp. 131-34.

8. *Ibid.,* pp. 135-39.

9. *Ibid.,* pp. 139-45.

10. J. Guintherius von Andernach, *De medicina veteri et noua tum cognoscenda, tum faciunda commentarij duo,* 2 vols. (Basel: Henricpetrina, 1571), *2,* p. 650.

11. *Ibid.,* pp. 26, 28, 621-22.

12. *Ibid.,* pp. 31-32.

13. I. Bernard Cohen's *Franklin and Newton: An Inquiry into Speculative Newtonian Experimental Science and Franklin's Work in Electricity as an Example Thereof* (Philadelphia: The American Philosophical Society, 1956) presents the argument that Newton's influence in the eighteenth century derived largely from his widely read *Opticks,* but there seems little doubt that the *Principia mathematica* was the foundation of his fame even if it had a smaller audience.

14. Arnold Thackray, *Atoms and Powers: An Essay on Newtonian Matter-Theory and the Development of Chemistry* (Cambridge: Harvard U.P., 1970), pp. 52-73.

15. John Freind, *The History of Physick; From the Time of Galen, to the beginning of the Sixteenth Century...,* 2 vols. 4th edition (London: M. Cooper, 1750), *2,* p. 204.

16. Nina Rattner Gelbart, «'Science' in Enlightenment Utopias: Power and Purpose in 18th Century French 'Voyages Imaginaires,'» (University of Chicago, Ph.D. dissertation, December 1973), p. 155 citing the *Journal de Bruxelles* (May 29, 1784), pp. 226-27.

17. *Ibid.,* p. 158.

18 Joseph Priestley, *The History and Present State of Discoveries Relating to Vision, Light and Colours* (London: J. Johnson, 1772); *The History and Present State of Electricity with Original Experiments* (London: J. Dodsley, J. Johnson and B. Davenport, 1767); Jean Étienne Montu-

cla, *Histoire de Mathématiques, dans laquelle on rend compte de leurs progrès... jusqu'à nos jours; où l'on expose le tableau et le développement des principales découvertes... et les principaux traits de la vie des mathématiciens les plus célèbres,* 2 vols. (Paris, 1758) expanded to 4 vols. (completed and edited by J. J. Le Francais de Lalande) (Paris: H. Agasse, 1799-1802; Jean Sylvain Bailly, *Histoire de l'Astronomie Ancienne, depuis son origine, jusquà l'éstablissement de l'école d'Alexandrie...* (Paris: Frères Dubure, 1775); *Histoire de l'Astronomie Moderne, depuis la fondation de l'école d'Alexandrie, jusqu'à l'époque de MDCCXXX,* 3 vols., (Paris: Les Frères de Bure, 1779-1782).

19. Hermann Boerhaave, «Prolegomena, or the History of Chemistry» in *A New Method of Chemistry; Including the Theory and Practice of that Art: Laid down on Mechanical Principles, and accommodated to the Uses of Life. The whole making a Clear and Rational System of Chemical Philosophy,* trans. P. Shaw and E. Chambers (London: J. Osborn and T. Longman, 1727), pp. 1-50. Pierre Simon de La Place, «Precis de l'Histoire de l'Astronomie» in *Exposition du Système du Monde,* 3rd ed., 2 vols. (Paris: Courcier, 1808), 2, pp. 259-415.

20. William Whewell, *History of the Inductive Sciences, From the Earliest to the Present Time,* 3rd ed., 2 vols. (New York: Appleton, 1873). For Whewell's approach see his «Preface» and «Introduction,» *1,* pp. 7-11, 41-51. He stated openly that «the work being aimed at being, not merely a narration of the facts in the history of science, but a basis for the Philosophy of Science» (p. 8).

21. *Ibid., 1,* pp. 185-239.

22. *Ibid., 2,* pp. 255.

23. The standard study is by Donald Fleming, *John William Draper and the Religion of Science* (Philadelphia: University of Pennsylvania Press, 1950; reprinted New York: Octagon Books, 1972).

24. John William Draper, *History of the Conflict Between Religion and Science,* 25th ed. (London: Kegan Paul, Trench, Trübner & Co., Ltd., 1910), pp. 350-51.

25. Andrew Dickson White, *A History of the Warfare of Science with Theology in Christendom,* 2 vols. (New York: Appleton, 1900), *1,* p. VIII.

26. *Ibid., p.* IX.

27. *James J. Walsh, The Popes and Science: The History of the Papal Relations to Science During the Middle Ages and Down to our Own Time* (Notre Dame Edition, New York: Fordham University Press, 1915), p. 334.

28. *Ibid.,* p. III.

29. *Ibid.,* pp. 500-19, Appendix IX, «The Danger of a Little Knowledge» (513).

30. A most interesting summary of the late nineteenth and early twentieth-century pioneers in the field (Moritz Cantor, Paul Tannery, Karl Sudhoff, Johan Ludvig Heiberg, Pierre Duhem, Sir Thomas Little Heath and Also Mieli) is to be found in George Sarton's «Acta atque Agenda» (1951) conveniently reprinted in *Sarton on the History of Science: Essays by George Sarton,* Selected and Edited by Dorothy Stimson (Cambridge: Harvard U.P., 1962), pp. 23-49.

31. George Sarton, *Introduction to the History of Science* 3 vols. in 5 (Baltimore: Published for the Carnegie Institution of Washington by Williams and Wilkins, 1927-1947), *1,* p. 3.

32. *Ibid.,* p. 6.

33. *Ibid.,* p. 19.

34. George Sarton, *A History of Science: Ancient Science Through the Golden Age of Greece* (Cambridge: Harvard University Press, 1952), p. XII.

35. *Ibid.,* p. XI.

LECTURE 2

THE HISTORY OF SCIENCE: PROFESSIONALIZATION AND DISUNITY

In my first lecture I spoke of the background to the establishment of the history of science as an academic discipline — ending with the almost monumental efforts of George Sarton pointing out that his positivistic approach to the field had much in common with the work of earlier historians. However, in contrast with his contemporaries and predecessors Sarton had founded journals and an international society for scholars in this field. At Harvard University his work had resulted in a program in the history of science both on the graduate and undergraduate levels. And in the 1940s this program was to graduate its first American Ph.D.s in the history of science.

It was not my intention to make these lectures autobiographical in nature, but I have found that to a certain extent it has been unavoidable because I am now speaking of a period when I first became interested in the history of science. I wrote a master's thesis on seventeenth-century chemistry at Indiana University in 1949 under John J. Murray, a professor of history who recognized the importance of the history of science. While doing research on that project I first became aware of *Isis* and the work of George Sarton. Then, after further graduate study in chemistry and five years of chemical research in the pharmaceutical industry my wife and I decided to return to graduate study. In 1956 there were only three graduate programs in the field in the United States: the Harvard Program directed by I. Bernard Cohen, the Cornell University program directed by Henry Guerlac, and the University of Winsconsin Program directed by Marshall Clagett. The first two were students of Sarton while the third had been trained by Lynn Thorndike, Columbia University's great medievalist who is best known for his eight volume *History of Magic and Experimental Science.*

Our decision was to go to Harvard where I enrolled in the autumn of 1956. George Sarton had died only a few months earlier. Announcements of his recent death had been sent to the subscribers of *Isis* were still being used as scratch paper in the Program Office at Widener Library. Not long after, many of his books were sold at library sales for as little as fifty cents each.

There were very few students in this somewhat esoteric field, but we soon found that the subject was being interpreted somewhat differently than we had expected. The most recent writings in the field were critical of Sarton and the author most frequently referred to as a model was Alexandre Koyré, the Russian philosopher of science who spent most of his later years in Paris. It is understandable that Koyré should have insisted on a close linkage between scientific and philosophical thought, but history was also important to him for only through it could we be given a sense of the «glorious progress» of the evolution of scientific ideas.[1] Like most other scholars in the field, Koyré centered his research on the development of physics and astronomy in the period from Copernicus to Newton. Galileo was an author of special concern, but he rejected the «Duhem thesis» — that is, that the sources for Galileo's mechanics were to be found in his medieval predecessors.[2] For Koyré, Galileo was an innovator far removed from the medieval critics of Aristotle, and if he had any predecessor at all, one would have to find him in Archimedes. He explained the Scientific Revolution as a fundamental change in world views (from Aristotelian to Copernican):

> ...I have endeavored in my *Galilean Studies* to define the structural patterns of the old and the new world-views and to determine the changes brought forth by the revolution of the seventeenth century. They seemed to me to be reducible to two fundamental and closely connected actions that I characterised as the destruction of the cosmos and the geometrization of space.[3]

This revolution was not to be explained by changes in society, a move from contemplation to active research, or even — he added — «the replacement of the teleological and organismic pattern of thinking and explanation by the mechanical and causal pattern.»[4] In many ways the Scientific Revolution was for Koyré the triumph of Plato over Aristotle in the Renaissance. And yet, if Sarton would have disagreed with Koyré over the importance of Plato for the rise of modern science, both would have agreed that the subject of the history of science was science and that this was the story of progress.[5]

As students we were also introduced to the encyclopedic works that in many ways characterized this «heroic age» of the history of science, Lynn Thorndike's *History of Magic and Experimental Science* published in eight volumes between 1923 and 1958[6]; Pierre Duhem's *Le Système du Monde,* ten volumes published between 1913 and 1959[7]; the many works of George Sarton, the first two (and what proved to be the only) volumes of Henry Sigerist's planned multi-volume history of medicine.[8] It seemed that vast areas of the field were just then being opened to our view. The multi-authored *A History of Technology* published in five volumes by Oxford gave us insight into a subject that had been ignored by most historians of science.[9] And in 1961 both the first volume of Joseph Needham's *Science and Civilisation in China,*[10] a work that is still underway and must rank as one of the great achievements of the present century, and the first volume of James Riddick Partington's *A History of Chemistry*[11] (actually volume two) were published. These works represented lifetimes of study by scholars who believed that they could cover entire fields over long chronological periods. It was a young field that had not yet reached the age of the specialized monograph.

However, it was also evident in the fifties that there were gaps in our learning. For those interested in Islamic science there did not seem to be very much to turn to. I had a special interest in the science of the Iberian peninsula, but other than accounts of the great voyages of discovery there was little to be found. A dedicated group of young scholars had already gathered round Otto Neugebauer and as a group they were rediscovering the mathematics and the astronomy of the ancient Near East. However, this group believed in specialization and they made little effort to integrate their research into the main stream of the history of science.[12]

Above all, the nineteenth century seemed to be a wasteland. Writing in 1954, I. Bernard Cohen noted that

> ...once we pass the boundary between the eighteenth and nineteenth centuries, we encounter no general surveys written in a way that will serve the historian of ideas. Merz's older work, dull, myopic, and often poorly written, is still the major presentation of the science of the nineteenth century. The fourth volume of Ernst Cassirer's *Erkenntnisproblem* is sketchy and far too technical for the average historian. Standard histories of physics, chemistry, biology, etc., contain much of the information but it needs to be digested and interpreted in the main stream of scientific ideas. Only the future can tell whether the history of science in the nineteenth century can be presented in a meaningful way for the general historian.[13]

Three years later Marshall Clagett gathered an international group of scholars at the University of Wisconsin to discuss current problems in the history of science. Published five years later, there is no better volume to indicate the state of the field a quarter century ago. The papers presented were heavily slanted toward the physical sciences and concentrate primarily on the period ranging from the late Middle Ages to the eighteenth century. In his preface to the collected papers Clagett commented that

> ...it might seem at first glance that we put too little emphasis on developments of the last century. The Committee would certainly agree that this is so. But I would stress that so few historians are doing serious and professional historical work in the history of science of the last few decades, that the presentation of a critical discussion of such problems would be

> most difficult. It might also seem that we slighted the biological developments in favor of those of the physical sciences. This was not our original intention. But our preliminary efforts to line up a notable group of people in the discussion of nineteenth-century biology was only partially successful. The field of those engaged in active research in biological history is so narrow that when we received some advance refusals, we had as a result to eliminate an additional day we had hoped to devote to biology.[14]

In fact, in the years since that meeting in Madison, historical research in nineteenth century science has far outstripped that in the period of the Scientific Revolution. However, this research has been somewhat uneven. Great strides have been taken in biological research which has centered on the history of evolutionary thought, but little has been done to synthesize the research on the history of the physical sciences.

Fully as important as the study of nineteenth century science has been the realization that the development of science may be influenced by factors we would not consider to be science at all. One of the first problems to arise related to Isaac Newton. Often praised as the greatest scientist of all time, his biographers frequently consciously ignored the fact that a large percentage of Newton's papers deal with alchemy and other matters which on the surface seem to have little to do with the foundation of classical physics and the establishment of the Copernican theory. Even more striking had been the neglect of Paracelsus, van Helmont, and their followers. Their work had been heatedly debated in the sixteenth and seventeenth centuries, but rejected as mystical (ergo, non-scientific) by the new Scientific Establishment of the late seventeenth century. Because of the positivistic bias of historians of science neither Newton's alchemy nor the mysticism of Paracelsus and van Helmont were «science». The mechanist philosophers of the Scientific Revolution had rightly excluded them from consideration and we should continue to do so.

George Sarton had dismissed alchemy, astrology and natural magic as «pseudo-science», but his decision to do this could rightly be questioned if historians of science ever chose a different approach to the field. Such a move was, indeed, underway among historians — especially English historians. In 1931 Herbert Butterfield published his influential essay, «The Whig Interpretation of History,» in which he argued that historians had, in effect, chosen sides. They had organized their histories from the viewpoint of the present, they clearly favored the Protestant reformers of the sixteenth and seventeenth centuries, and they defined «progress» from that viewpoint. In political terms they were guilty of writing «Whiggish» history. These historians felt the need to give a verdict, and in so doing they oversimplified the rich complexity of their sources. He wrote that

> The value of history lies in the richness of its recovery of the concrete life

> of the past. It is a story that cannot be told in dry lines, and its meaning cannot be conveyed in a species of geometry. There is not an essence of history that can be got by evaporating the human and the personal factors, the incidental or momentary of local things, and the circumstantial elements, as though at the bottom of the well there were something absolute, some truth independent of time or circumstance.... [15] Above all it is not the role of the historian to come to what might be called judgements of value.... [16] His role is to describe; he stands impartial between Christian and Mohammedan; he is interested in neither one religion nor the other except as they are entangled in human lives.... [17] He is back in his proper place when he takes us away from simple and absolute judgements and by returning to the historical context entangles everything up again. He is back in his proper place when he tells us that a thing is good or harmful according to circumstances, according to the interactions that are produced. If history can do anything it is to remind us of those complications that undermine our certainties, and to show us that all our judgements are merely relative to time and circumstance. [18]

Butterfield's «manifesto» was a challenge to all historians. In fact, he was later to become interested specifically in the history of science and we shall return to him shortly.

Among historians of science and medicine, Walter Pagel was one of the first to draw attention to the neglected figures of history. [19] But although his first book on van Helmont appeared in 1930, his widespread methodological influence is more recent, dating from the publication of his *Paracelsus* (1958) and his *William Harvey's Biological Ideas* (1967). Recognizing the fallacy of Sarton's «history of the gradual revelation of truth,» Pagel has countered that such an approach «based on the selection of material from the modern point of view, may endanger the presentation of historical truth.» [20] Indeed, histories in which «discoveries and theories of the past are taken from their original context to be judged alongside modern scientific and medical entities» are likely to be dangerously misleading. [21]

How then should the historian of science proceed? Referring to his own research, Pagel has suggested that:

> Instead of selecting data that «make sense» to the acolyte of modern science, the historian should therefore try to make sense of the philosophical, mystical or religious «side-steps» of otherwise «sound» scientific workers of the past — «side-steps» that are usually excused by the spirit or rather backwardness of the period. It is these that present a challenge to the historian: to uncover the internal reason and justification for their presence in the mind of the savant and their organic coherence with his scientific ideas. In other words it is for the historian to reverse the method of

> scientific selection and to restate the thoughts of his hero in their original setting. The two sets of thought — the scientific and the non-scientific — will then emerge not as simply juxtaposed or as having been conceived in spite of one another, but as an organic whole in which they support and confirm each other. There is no other way to lay the savant open to our understanding.[22]

It has thus been Walter Pagel's desire to interpret the facts of medical and scientific history «as the outward expression of their time.» When this has been done, he explains,

> It will then appear that not only certain standards of technical equipment made discoveries possible, but that these can be seen also as the offspring of certain non-scientific ideas and of a particular cultural background. ...The History of Medicine will then appear much more complicated than it does in the usual perspective of straight lines of progress. Yet we will have to embark on the cumbersome task of reconstructing ancient thought if we wish to write history — and not best sellers.[23]

Pagel once told me that after listening to Koyré lecture on Isaac Newton's physics he rose to ask about Newton's work on alchemy. Koyré dismissed the point by saying that «we are not concerned with that.» From his point of view he was right, but for Pagel it is impossible to understand the «total man» unless we examine *all* of his work. Perhaps nothing better illustrates the basic differences between these two scholars than this anecdote.

Important though Pagel's work is, his influence may well have been less than that of the late Dame Frances Yates who wrote a series of books relating the Scientific Revolution to Hermeticism. A literary historian, Dame Frances first attracted the attention of historians of science with the publication of her *Giordano Bruno and the Hermetic Tradition* in 1964.[24] Here was an attempt to assess the work of Bruno as a sixteenth-century supporter of the heliocentric theory not because he was a forward-looking scientist, but because the sun-centered system best accommodated his mystical, «Hermetic,» views of the sun and the universe. This book is surely one of the most influential to have been published in the history of science in the past two decades. And on the whole this influence has been healthy, since she urged historians to cope with a vast body of texts that never should have been ignored in the past.

The Yates influence has also had dangerous side effects. Overwhelmed by the importance of Hermeticism, Neo-Platonism, magic and other mystical strands of Renaissance philosophy, Frances Yates went on to ever more daring positions that were based upon less and less solid evidence. In *Rosicrucian Enlightenment* (1972), she came close to insisting that the entire Scientific Revolution developed from Renaissance mysticism and magic.[25] Here she strove to connect the origins of the Royal Society of London as well as the work of Descartes and Newton with John Dee and the Rosicrucian documents of the early years of the century.[26] Unfortunately, these suggestions have not been upheld by the sound historical evidence they require. They are speculations which must be considered doubtful at best.

The works of Pagel and Yates have generated considerable interest and debate. And it may be worth noting that neither of them represents any of the older, traditional fields in the history of science. Yates may best be described as a literary historian while Pagel was a classicist and a pathologist as well as an historian of medicine. But they both offer a challenging approach to the history of science — one that may go far toward solving the entire problem of the Scientific Revolution.

Among historians of science the study of the pseudo-sciences has aroused the most conflict in relation to the proper interpretation of the work of Isaac Newton. All would agree that Newton represents the culmination of many strands of early modern physics, mathematics, and astronomy, but how is one to interpret the thousands of folios of alchemical manuscripts he wrote? Among the first to try to integrate them into a total picture of Newton has been R. S. Westfall who had earlier discussed this author only in terms of the traditional internal history of science. By the early 1970s Westfall had become convinced that the Hermetic mysticism of the seventeenth century was an essential ingredient in Newton's thought and that this «could lead the relatively crude mechanical philosophy of the seventeenth-century science to a higher plane of sophistication.»

> The Hermetic elements in Newton's thought were not in the end antithetical to the scientific enterprise. Quite the contrary, by wedding the two traditions, the Hermetic and the mechanical, to each other, he established the family line that claims as its direct descendants the very science that sneers today uncomprehendingly at the occult ideas associated with Hermetic philosophy.[27]

And, in the most recent contribution to the study of Newton's alchemy, B. J. T. Dobbs goes further to claim not only that much of Newton's most important work derives from his alchemical speculations, but that «in a sense the whole of his career after 1675 may be seen as one long attempt to integrate alchemy and the mechanical philosophy.»[28]

It is little wonder that more traditional historians of science have expressed their fears of these new developments. At a meeting at King's College, Cambridge (1968) devoted to new trends in the field, P. M. Rattansi argued the case for contextual history stating that the «historians' task cannot be that of isolating 'rational' and 'irrational' components, but of regarding it as a unity and locating points of conflict and tension only on the basis of an exploration in considerable depth.» [29] In reply, Mary Hesse argued against the inclusion in the field of subjects that were not truly scientific in modern terms. The pseudo-sciences might well belong to history, but they could not be considered as part of the history of science. Anticipating that such an approach might be considered exclusive, she added that it is essential that we should use modern science as a means of weighing the arguments of the past. To use judgements of the past that include non-scientific elements is a waste of our time. Indeed, she concluded, we must be careful what we read or permit ourselves to assess since, by «throwing more light on a picture, we may distort what has already been seen.» [30] Hesse's reaction is by far the most extreme yet to surface from the more traditional historians and philosophers of science.

The impasse evident in the exchange between Hesse and Rattansi exhibits the tension currently existing in the field. And yet, the role of the so-called pseudo-sciences is hardly the only source for this. Perhaps the sharpest debate at the moment concerns the relation of science and society. Only a few years ago this seemed to be relatively unimportant. When Thomas S. Kuhn prepared a history of science for the *Encyclopedia of the Social Sciences* (1968) he compared «internalist» with «externalist» histories of science. The former dealt with technical questions related to the growth of science; the latter were «attempts to set science in a cultural context which might enhance understanding both of its development and its effects....» [31] Of special interest was the debate over the thesis proposed by Thomas K. Merton (1938) which sought to explain the success of seventeenth-century science in England by pointing to (a) the Baconian emphasis on practical arts and trade processes and (b) the stimulus of Puritanism in religion. [32] But if Kuhn argued that internal and external histories of science are complementary, he also felt that this was an older argument made for the most part by scholars who had not succeeded in proving their point. Referring to the «new generation of historians» — that is, for the most part the Koyré inspired historians — they

> claim to have shown that the radical sixteenth- and seventeenth-century revisions of astronomy, mathematics, mechanics, and even optics owed very litlle to new instruments, experiments, or observations. Galileo's primary method, they argue, was the traditional thought experiment of scholastic science brought to a new perfection. [33]

This was far removed from the craft tradition or the new methodology of Bacon which failed consistently. As far as the seventeenth-century is concerned, he suggested that only the «new» sciences such as electricity and magnetism, chemistry, and thermal phenomena borrowed from the craft tradition. [34] The mathematical sciences should continue to be studied by internal methods.

Kuhn's much lauded *Structure of Scientific Revolutions* (1952) is an internalist study seeking to explain scientific revolution in terms of the replacement of one scientific paradigm with another.[35] With ever increasing interest in non-scientific factors in the growth of science this work has not strongly affected historians of science as much as one might expect. Rather, it has appealed most to social scientists, philosophers and historians who have used it less as model for the history of science than to examine the internal development of their own fields.[36]

For Thomas Kuhn the «new» history of science was to be primarily internalist. However, the late sixties and the early seventies were to see an ever-growing interest in the interrelation of science with society. For this reason the history of science has become a field far more attractive to historians, philosophers and social scientists — many of whom have had little training in either the sciences or the history of science. These authors argue that significant aspects of scientific history may now be grasped without the technical scientific knowledge that earlier seemed to be essential. There have been mixed results from this since, in fact, technical knowledge does remain important even in some of the most esoteric areas of the history of science. Still, a number of important studies have appeared. For example, Keith Thomas' *Religion and the Decline of Magic* (1971) is a monumental contribution to our understanding of the early modern intellectual scene in England.[37] No less important is the work of Christopher Hill who has used the recent studies of alchemy and the Paracelsians as an integral key to his understanding of the Civil War in England in his *The World Turned Upside Down* (1972).[38]

In *The Newtonians and the English Revolution, 1689-1720* (1976), Margaret Jacob has argued that the triumph of Newtonian physics may have been due less to the value of Newton's science than it was to the fact that English theologians in the period of the «Glorious Revolution» (1688) sought a powerful ally through their espousal of the Newtonian synthesis.

> These churchmen... used the new mechanical philosophy in support of Christianity and in their assault on atheism and thus spread the ideas of the new science and its concomitant natural philosophy. Without the sermons of the first latitudinarians, science would have remained esoteric and possibly even feared by the educated but pious public.[39]

She sees the new science as an explicit rejection of the older mechanical philosophies of Hobbes and Descartes as well as the Aristotelianism of the universities and the radical cosmologies of the mid-century which had too often been associated with those who rose in rebellion against the Church and State.

> The obvious question of course, is why these liberal churchmen felt impelled to reject one science of natural philosophy and accept another. Historians of science have often presumed that the new mechanical philosophy triumphed in England simply because it offered the most plausible

> explanation of nature. It may do just that, but in my understanding of the historical process that made it acceptable the supposed correspondence of the new mechanical philosophy with the actual behavior of the natural order is not the primary reason for its early success.[40]

For Jacob, the social explanation of the triumph of Newtonianism is to be found in «its usefulness to the intellectual leaders of the Anglican Church as an underpinning for their vision of what they liked to call the 'world politick.' The ordered, providentially guided, mathematically regulated universe of Newton gave a model for a stable and prosperous polity, ruled by the self-interest of men.»[41] In short, we see here an explanation of the Newtonian triumph on grounds totally divorced from the fact that Newton's work represents the culmination of nearly a century and a half of scientific discussions and research leading from the *De revolutionibus orbium* (1543) of Copernicus to the *Principia mathematica* (1687).

At a meeting of the American Association for the Advancement of Science held in December 1979, Charles C. Gillispie lashed out against those who followed newer trends in the field. As reported in *Science* Gillispie complained that «the history of science is losing its grip on science, leaning heavily on social history, and dabbling with shoddy scholarship. He attacked those who discussed scientific problems but who had little or no scientific training.

> Less odious but still troublesome to Gillispie are social histories that ignore science altogether, such as studies that deal with the role of women in a particular scientific institution but omit their actual scientific work.... Another trend, he said, is that scholars focus on the personal and anecdotal: Newton on alchemy rather than on motion, Kekule's snake dance rather than the benzene ring, Darwin's neurosis rather than his marshaling of evidence. Some so-called scholars focus on scandal.... «These scholars,» says Gillispie, «have a lust for just the sort of thing most rigidly ruled out of court in the science we do now — the irrational, the personal.»[42]

Gillispie's plea for a return to the values of Koyré has been dismissed by the social historians who have replied that

> The social history of science has by now established itself within the discipline as a legitimate method of approaching the past. Despite recent rearguard action, notably by C. C. Gillispie, most historians accept that the traditional practices of analyzing theoretical developments within the sciences need to be supplemented by the study of the changing social foundations of scientific activity. The «internal vs. external» debates of the late 1960s are, one hopes a thing of the past.[43]

A similar debate may be seen currently in the history of medicine where it used to be considered essential for the scholar to take an M.D. prior to beginning work on a second doctorate in the history of medicine. His research was then expected to center on scientific aspects of medical theory and disease. This tradition continues, but it is no longer in the ascendant. Today the cutting edge of medical history is centered on the social setting of medicine. These historians are most frequently being trained through programs located in departments of history rather than the older departments or institutes of the history of medicine. And since these scholars do not have the medical training of their predecessors we are witnessing debates in the learned journals relating to the subject matter and the proper training for the history of medicine.[44]

At the death of George Sarton the history of science was established as a small field, but one that was recognized by many as having importance. However, because of its historical development it was to be found in the academic world most frequently in the form of programs independent of history or science. Most publishing historians of science twenty-five years ago had been trained as scientists. Sarton recognized this, but believed that in the future the professional historian of science should have at least two masters' degree — one in a science and the other in history — before proceeding on to his Ph.D. in the history of science. However, the influence of Koyré and a trend among philosophers away from the history of philosophy toward the philosophy of science emphasized the growth of independent programs in the history *and* philosophy of science. During the fifties and the sixties there were further discussions of the relationship of the history of science to both history and the sciences.

In 1956 it seemed clear that the history of science required an expertise in the sciences that seemed to set it apart from the training received by all but the most unusual historians. But at this time traditional historians were becoming aware of the tremendous impact of science and technology on our lives and this gave rise to a certain urgency to learn more of this field. Thus, in a lecture on «The History of Science and the Study of History» in 1959 Herbert Butterfield said that,

> Although the world had long known that science and technology were important, it is only recently that these things have taken command of our destiny — that destiny which we had learned from our history books to regard as depending so greatly on the wills of statesmen.[45]

He argued that historians must take into account the rise of modern science and that when they do this it will «change the whole character of historiography.»[46] And yet Butterfield did not challenge the independence of the history of science. In his influential *The Origins of Modern Science 1300-1800* (1949), he presented the customary positivistic approach to the field that was current in the immediate post-war years.[47] The history of science must be understood by historians, but the field could rightly develop on its own because of the specialized knowledge it required. In fact Butterfield's call for a greater awareness of science by historians was heeded. As more and more doctorates were awarded in the history of science in the 1960s and the 1970s most of these young scholars found themselves hired by departments of history rather than by the older independent programs in the history of science or the history and philosophy of science. This new interest among traditional historians surely accelerated the move toward new areas of research such as those I have already noted, the part played by the pseudo-sciences in the rise of modern science as well as more general topics relating science to society and culture. The development of the field in recent decades has also reopened the question of the relationship of the history of science to the sciences. In the 1950s few scientists were more influential in arguing reform in scientific education than James B. Conant. The War had shown clearly the need for more advanced scientific training for American youth. As a result, the method of teaching the sciences was rethought — and, at the same time, historical «case-studies» were introduced to give non-science majoring undergraduates the opportunity to see how the sciences have developed. But Conant, in a lecture delivered in 1960, stated that history was just as valuable for the scientist. He believed that scientific education was frequently too narrow and that the use of the case history approach would give students vision that would be broader and more informed.[48] He outlined a new science curriculum that would prepare students first in the history of their own science speciality and then in the history of modern science. These courses were to be followed by others in the history of science taken in its widest possible sense and — only then — cultural and political history which would be understood in connection with the earlier courses in the history of science. Because of its value for the sciences Conant argued for strong history of science departments everywhere. He was far less enthusiastic about those who sought to equate the history of science with the social history of science — or the philosophy of science.[49]

Conant's ambitious plans for the historical training of scientists never bore fruit although the case-history approach to science was extensively employed in American universities in the fifties and sixties. However, eventually these courses were largely abandoned. For scientists they seemed to move too slowly for the mass of material required to be taught. For historians of science they were not historical enough — and for students with little interest in the sciences these courses were often less interesting than a concentrated survey course in a particular science might have been.[50]

Although there are several notable cases in the history of science in which a scientist's knowledge of much older literature led to a breakthrough, these cases are few and far between. I can recall a discussion presided over by Herbert Butterfield in 1959. There were some thirty students and faculty members present and Butterfield asked if anyone knew of a case in which a knowledge of the history of science proved to be of direct value in a scientific discovery. I was the only one present who answered — and that was only possible because I had then recently been working on the discovery of the inert gases and knew of Ramsay's reading of the eighteenth-century papers by Cavendish and of their influence on him.

Thomas Kuhn agreed that the sciences might not benefit much from the reading of the history of science. In 1968 he wrote that

> Among the areas to which the history of science relates, the one least likely to be significantly affected is scientific research itself. Advocates of the history of science have occasionally described their field as a rich repository of forgotten ideas and methods, a few of which might well dissolve contemporary scientific dilemmas. When a new concept or theory is successfully deployed in a science, some previously ignored precedent is usually discovered in the earlier literature in the field. It is natural to wonder whether attention to history might not have accelerated the innovation. Almost certainly, however, the answer is no. The quantity of material to be searched, the absence of appropriate indexing categories, and the subtle but usually vast differences between the anticipation and the effective innovation, all combine to suggest the reinvention rather than rediscovery will remain the most efficient source of scientific novelty.[51]

In general I would agree with Kuhn's conclusion: ...we should not study the history of science only in the hope of finding long forgotten laws of discoveries still valid today. And yet, the history of science does have a real value for scientists as well as non-scientists. Through a study of the earlier literature in his own field the student will become aware of the scientific process. He should surely learn to appreciate the fact that the results we accept today were not generally arrived at in a simple manner — and that the same process of debate is part of science today. I have heard historians of astronomy argue that a knowledge of the history of astronomy does help to clarify current astronomical debates. And we find historians of Darwinian evolution active both in the biological sciences and in the heated Creationism debates.

During the past hour I have chosen to illustrate the change in this rapidly growing field by concentrating on only a few factors, its growth in subject matter, the debates that have occurred over the introduction of the so-called pseudo-sciences and societal factors in the growth of modern science, and the discussion of the relationship of the history of science to both history and the sciences. I might well have turned to other topics such as the new interest in the history of technology and its relationship to scientific research or the study of science in specific national settings, but I believe that the examples I have chosen will suffice to indicate the types of problems that have engaged the members of the profession over the past few decades.

I sometimes wonder what George Sarton would think of the field if he were still alive. At the time of his death his positivistic views were still dominant. Today they are passé. The history of science has developed to its present state in a relatively short period of time, but this maturity has also been accompanied by a decay of the comfortable sense of progress with which it was once identified. Indeed, the present defensive attitude of some historians of science indicates just how far the interest in new methodologies has penetrated the field. Today in addition to scientists and historians of science there are a number of historians, literary critics, and social scientists who are effectively applying the history of science to their own areas of research.

And yet there is really no need to fear — as Gillispie evidently does — that the history of science will lose its need for the technical data of the sciences. The history of science will always require technical research in the sciences and the histories so produced will always have value. But this does not mean that we should be unwilling to go beyond the technical monographic studies and the technical criticism that have characterized much of the post-war scholarship in this field. There must be room for syntheses that lead to a broader understanding of science as a whole as well as its interrelation with other areas of human endeavor. I think that we should learn to apply the method of Walter Pagel to understand discoveries in terms of the entire work of the discoverer and then go on to understand the discoverer in terms of the total intellectual milieu that shaped him. Only when this is done will it be possible for our histories to truly reflect the impact of science on civilization. Only when this is done will we see that there is no real conflict between the traditional internalist and externalist schools of the history of science. And only when this is done will the value of the history of science be universally accepted by historians and scientists alike.

NOTES

1. Alexandre Koyré, *Études Galiléennes,* 3 parts, 1935-1939; reprinted in one volume (Paris: Hermann, 1966), p. 11.

2. See especially Pierre Duhem, *Le Système du Monde,* 10 vols. (Paris: Hermann, 1913-1959).

3. Alexandre Koyré, *From the Closed World to the Infinite Universe* (New York: Harper Torchbook, 1958), p. VI.

4. *Ibid.,* p. v.

5. For Sarton on Plato see *A History of Science Through the Golden Age of Greece* (Cambridge: Harvard U.P., 1952), pp. 395-430. He discussed Plato's scientific thought in relation to the *Timaios* and wrote that

   «The influence of *Timaios* upon later times was enormous and essentially evil.... The scientific perversities of *Timaios* were mistaken for scientific truth. I cannot mention any other work whose influence was more mischievous, except perhaps the Revelation of St. John the Divine. The apocalypse, however, was accepted as a religious book, the *Timaios* as a scientific one; errors and superstitions are never more dangerous than when they are offered to us under the cloak of science.»

6. Lynn Thorndike, *A History of Magic and Experimental Science,* 8 vols. (New York: Columbia University Press, 1923-1958).

7. See above, note 2.

8. Henry Sigerist, *A History of Medicine,* 2 vols. (New York: Oxford U.P., 1955, 1961).

9. Charles Singer, E. J. Holmyard and A. R. Hall, eds., *A History of Technology,* 5 vols. (New York and London: Oxford U.P., 1954-1958).

10. Joseph Needham, *Science and Civilisation in China, Volume 1: Introductory Orientations* (Cambridge: Cambridge U.P., 1961).

11. The first volume published was the second covering the sixteenth and the seventeenth centuries. J. R. Partington, *A History of Chemistry* 2 (London: Macmillan, 1961).

12. «I am exceedingly sceptical of any attempt to reach a 'synthesis' — whatever this term may mean — and I am conviced that specialization is the only basis of sound knowledge.» O. Neugebauer, *The Exact Sciences in Antiquity* (1952; New York: Harper Torchbooks, 1962), pp. V-VI.

13. I. Bernard Cohen, «Some Recent Books on the History of Science,» *Roots of Scientific Thought: A Cultural Perspective* eds. Philip P. Wiener and Aaron Noland (New York: Basic Books, 1957), pp. 627-56 (656). Published originally in the *Journal of the History of Ideas.*

14. *Marshall Clagett, ed. Critical Problems in the History of Science: Proceedings of the Institute for the History of Science at the University of Wisconsin, September 1-11, 1957 (Madison: The University of Wisconsin Press, 1962), p.* VI.

15. *Herbert Butterfield, The Whig Interpretation of History* (1931; New York: W. W. Norton & Co. Inc., 1965), p. 68.

16. *Ibid.,* p. 73.

17. *Ibid.,* p. 74.

18. *Ibid.*, p. 74-75. For a recent discussion of Butterfield and «Whiggish» history see A. Rupert Hall, «On Whiggism,» *History of Science 21* (1983), pp. 45-59.

19. Walter Pagel, *Jo. Bapt. Van Helmont: Einführung in die philosophische Medizin des Barock* (Berlin: Springer, 1930); *Paracelsus: An Introduction to Philosophical Medicine in the Era of the Renaissance* (Basel/New York: S. Karger, 1958); *William Harvey's Biological Ideas: Selected Aspects and Historical Background* (Basel/New York: S. Karger, 1967).

20. Walter Pagel, «The Vindication of Rubbish,» *Middlesex Hospital Journal* (Autumn, 1945), pp. 1-4 (1).

21. *Ibid.*

22. *Pagel, Harvey's Biological Ideas,* p. 82.

23. Pagel, «Vindication of Rubbish,» p.4.

24. Frances A. Yates, *Giordano Bruno and the Hermetic Tradition* (Chicago: The University of Chicago Press; London: Routledge & Kegan Paul; Toronto: The University of Toronto Press, 1964).

25. Frances A. Yates, *The Rosicrucian Enlightenment* (London and Boston: Routledge & Kegan Paul, 1972).

26. See *ibid., pp. 113, 171-205.*

27. *Richard S. Westfall, «Newton and the Hermetic Tradition» in Allen G. Debus, ed. Science, Medicine and Society in the Renaissance: Essays to honor Walter Pagel,* 2 vols. (New York: Science History Publications, 1972) *2,* pp. 183-198 (195).

28. Betty Jo Teeter Dobbs, *The Foundations of Newton's Alchemy or «The Hunting of the Greene Lyon»* (Cambridge/London/New York/Melbourne: Cambridge U.P., 1975), p. 230.

29. P. M. Rattansi, «Some Evaluations of Reason in Sixteenth and Seventeenth-Century Natural Philosophy,» *Changing Perspectives in the History of Science: Essays in Hornour of Joseph Needham,* eds. Mikulás Teich and Robert Young (London: Heinemann, 1973), pp. 148-66 (150).

30. Mary Hesse, «Reasons and Evaluations in the History of Science,» *ibid.,* pp. 127-147 (143).

31. Thomas S. Kuhn, «History of Science,» *International Encyclopedia of Social Sciences,* ed. David L. Sills (New York: Crowell Collier and Macmillan, 1968, 1979) *14, pp. 75-83 (76-82).*

32. *Ibid., pp. 79-81.*

33. *Ibid.,* p. 80.

34. *Ibid.*

35. *Thomas S. Kuhn, The Structure of Scientific Revolutions* (Chicago: The University of Chicago Press, 1962). This book was also issued as volume 2, number 2 of the *International Encycolpedia of Unified Science* published by The University of Chicago Press.

36. As examples of this literature see the following: Barry Barnes, *T. S. Kuhn and Social Sciences* (New York: Columbia University Press, 1982); Signe Seiler, *Wissenshaftstheorie in der Ethnologie: zur Kritik u. Weiterführung d. Theorie von Thomas S. Kuhn anhand ethnograph* (Berlin: Reimer, 1980); Garry Gutting, ed. *Paradigms and Revolutions: Appraisals and Applications of Thomas Kuhn's Philosophy of Science* (Notre Dame: University of Notre Dame Press, c. 1980).

37. Keith Thomas, *Religion and the Decline of Magic: Studies in Popular Beliefs in Sixteenth- and Seventeenth-Century England (1971; Harmondsworth: Penguin, 1973).*

38. *Christopher Hill, The World Turned Upside Down: Radical Ideas During the English Revolution* (1972; New York: The Viking Press, 1973), see especially pp. 231-46.

39. Margaret C. Jacob, *The Newtonians and the English Revolution 1689-1720* (Ithaca: Cornell U.P., 1976), pp. 16-17.

40. *Ibid.,* p. 17.

41. *Ibid.,* p. 18.

42. William J. Broad, «History of Science Losing Its Science,» *Science 207* (January 25, 1980), 389.

43. Paul Wood, «Recent Trends in the History of Science: The dehumanisation of history,» *BSHS Newsletter,* No. 3 (September, 1980), 19-20 (19).

44. Here see Leonard G. Wilson, «Medical History Without Medicine,» *Journal of the History of Medicine and Allied Sciences 35* (1980), 5-7; Lloyd G. Stevenson, «A Second Opinion,» *Bulletin of the History of Medicine 54* (1980), 135; Ronald L. Numbers, «The History of American Medicine: A Field in Ferment,» *Reviews in American History 10* (1982), 245-64; Gert H. Brieger, «The History of Medicine and the History of Science,» *Isis 72* (1981), 537-40.

45. Herbert Butterfield, «The History of Science and the Study of History,» *Harvard Library Bulletin 13* (1959), 329-47 (329).

46. *Ibid.,* p. 347.

47. Herbert Butterfield, *The Origins of Modern Science 1300-1800* (New York: Macmillan, 1952).

48. James B. Conant, «History in the Education of Scientists,» *Harvard Library Bulletin 14* (1960), 315-33 (322-23).

49. *Ibid.,* 325.

50. This assessment is my own after having taught courses of this genre for four years both at Harvard University and the University of Chicago during the years 1957-59 and 1961-63.

51. Kuhn, «History of Science,» p. 81.

LECTURE 3

THE SIGNIFICANCE OF CHEMICAL HISTORY

The history of the writing of the history of chemistry is in some ways an enigma to its students. A practicing chemist myself, I had learned my chemistry from textbooks which invariably began with historical introductions. The departments in which I had studied had offered courses in the history of chemistry at a time when this was not unusual.[1] These courses may have drawn few students, but certainly the field seemed to be a well established one. However, when I returned to graduate work at Harvard expecting to study the history of chemistry I found that I was something of a rebel. There the history of chemistry was ignored by the historians of science. In the survey course on the history of science little more was referred to than Lavoisier while in the graduate level seminar on the history of the physical sciences chemistry was totally ignored. At the time this surprised me, but the answer is clear if one turns to history.

As a student in chemistry I had been correct in thinking that there has been a strong historical tradition in this science. The custom may be less common today, but most twentieth-century authors of chemical textbooks who felt compelled to begin their works with some sort of historical introduction were maintaining a custom that may be traced back hundreds of years. I first became acquainted with the history of chemistry through the use of Partington's *Text-Book of Inorganic Chemistry*[2] and Sneed and Maynard's *General Inorganic Chemistry*,[3] both of which have introductory historical sections. At the end of the last century the still useful *Treatise on Chemistry* by Roscoe and Schorlemmer[4] began with a forty-page history, while the names of Ramsay, Berthelot, Kopp and Schorlemmer bring to mind authorative historical as well as chemical achievements.[5] The custom of introducing any general work on chemistry with a history of that science was so common by the beginning of the nineteenth century that, when William Henry proposed to write a text without this standard feature, he felt it necessary to explain why he had omitted it.[6] Joseph Black's late eighteenth-century lectures on chemistry are famous for indicating his early acceptance of the chemistry of Lavoisier.[7] In published form they lack a history of science, but working from manuscript copies of the lectures, Douglas McKie showed that they, too, began with an historical introduction. Any number of examples from the time of Black to the present might be given to show this persistent interest in the history of chemistry.

For the period prior to Black one thinks immediately of Boerhaave's introductory survey of the development of the subject, while Black's lectures seem to have been strongly influenced by the historical debate between Hermann Conringius (1606-1681) and Olaus Borrichius (1626-1690). Conringius, one of the most learned classics of the day, taught at Helmstedt and published his *De hermetica Aegyptiorum vetere et Paracelsicorum nova Medicina* in 1648. An Aristotelian and a Galenist, he opposed the chemical medicines of the Paracelsians and rejected the historical existence of Hermes Trismegistus. Borrichius, a professor of chemistry and botany at Copenhagen, attacked this work twenty years later in his *De ortu et progressu chemiae* (1668). Here he pointed to evidence that the origins of chemical knowledge could be traced to a very early date, at least to Tubal Cain long before the Deluge. As for Hermes, Borrichius had no doubt whatsoever. Not only did chemistry originate in Egypt, medicine did as well and the person who originated both was Hermes who was the Greek Mercury.[8] In short, Conringius was mistaken regarding the antiquity of the art and the role of Hermes Trismegistus in its transmission to the rest of the world from Egypt.

Conringius, respected as one of the most learned scholars of Europe, was deeply offended by this work of Borrichius and he published a new and greatly expanded version of his *De hermetica medicina* in 1669 to which he appended an «Apologeticus adversus calumnias et insectationes Olai Borrichii.» Here he argued that it was impossible to know the original Hermetic doctrines because they had been so altered by time, but what we do know is highly suspect. Hermetic medicine is infected with magic and impiety... indeed, Egyptian knowledge as a whole was impious and filled with superstitious rites. Those who write of the importance of Egyptian magic are themselves suspect since this magic is not natural but demonic magic. As for Paracelsus and his followers, their work has corrupted all philosophy. Their three principles are useless and their metallic medicines have been plagiarized from earlier physicians such as Arnold of Villanova and Raymond Lull. As for the rest, it is «partim impiam... partim vanam et absurdum.»[9]

Borrichius did not let the case rest at this point. In a new work, the *Hermetis, Aegyptiorum, et chemicorum sapientia ab Hermanni Conringii animadversionibus vindicata per Olaum Borrichium* (1674) he replied that Hermes was most certainly an historical figure and that the art of transmutation had been discovered by him. This was the reason why the Egyptians had been able to accumulate enough wealth to carry out their vast building projects. Nor were the works of the later Greeks — such as Aristotle, Theophrastus, Euclid and Ptolemy — as perfect as Conringius would wish. And what was to be praised in them might well be ascribed to the fact that many of them had studied in Egypt and had partaken of the ancient wisdom still being taught there.

> And as for *Hippocrates* and *Galen,... Cos,* the Country of the former, was so near *Aegypt,* that doubtless he thence received great advantage to his Medical knowledge; and that *Democritus,* his Master, who had been long acquainted with *Aegypt,* had questionless suggested many things to him: That *Galen* also had lived long at *Alexandria,* and was wont to advise the Grecian Candidates of Physick to travel thither for experience. As for *Ptolemy,* that he was no Grecian, but an *Alexandrian,* or a *Pelusiot,* and consequently of *Aegypt.* [10]

Borrichius attacked Conringius further for his rejection of the value of chemical medicines. As for Paracelsus, he was pictured as a great master who had rediscovered the truths known to the ancient Egiptian adepts.

The conflict between Borrichius and Conringius was given wide publicity through lengthy reviews in the *Philosophical Transactions* of the Royal Society of London in 1668, 1674, and in the *Journal des Scavans in 1675.* [11] The points discussed regarding the antiquity of chemistry were considered of sufficient significance for the major texts the *De hermetica medicina of Conringius* and the *De ortu et progressu of Borrichius,* to be considered fundamental sources for chemical historians until the close of the eighteenth century. [12] However, the question for us to answer is why the antiquity of chemistry should stir up such a heated debate. If we look at the French chemical textbooks of the seventeenth century we find that their authors, men such as Beguin, Lefèvre, and Lemery paid scant attention to the historical background of their subject, some none at all. [13] However, if we turn to the Paracelsians and alchemists of this period we see that the historical interest is derived from the alchemical-iatrochemical tradition rather than from the practical texts. Peter Severinus began his Paracelsian apology, the *Idea medicinae philosophicae* (1571) with a chapter on the origin and progress of the medical art, the main purpose of which was to emphasize the importance of Paracelsus and his application of chemistry and medicine. [14] Early in the next century Daniel Sennert was profoundly influenced by the iatrochemical movement, and in his *De chymicorum cum Aristotelicis et Galenicis consensu as dissensu* (1619) he included a chapter that examined the origins of chemistry. Here he questioned the claims of some that Adam was the first alchemist, but he willingly accepted Tubal Cain as the most probable originator of the rudiments of the art. [15] With Sennert we are already in contact with the traditional alchemical works, and if we examine the two major collected editions of alchemical tracts, we see that they — like Joseph Black's lectures or many modern chemical textbooks — begin with introductory historical studies. The first volume of Lazarus

Zetzner's *Theatrum Chemicum* (1602) opens with the *De veritate & antiquitate artis chemicae...* of Robertus Vallensis,[16] while J. J. Manget's *Bibliotheca Chemica Curiosa* (1702) begins with a complete reprint of the *De ortu & progressu chemiae* of Olaus Borrichius.[17]

A fascination with the antiquity of their art was natural to the alchemists who sought to connect their knowledge with the divine truths known to Adam prior to the Fall. To give their writings greater authority these men often ascribed their work to legendary personages of the dim past — a custom that remains a perpetual headache for modern historians. Thus from the earliest times alchemists endeavored to establish the antiquity of their learning, and it is on the basis of this alchemical tradition that the standard alchemical and iatrochemical histories of the subject were founded. These were the «extravagant» histories which Joseph Black called «very wild & absurd,» yet which he used as the basis of the early period in his own lectures.[18] But in reality Black's history forms but one chapter of a long tradition, and in a real sense one may connect modern scholarship in this field with the historian alchemists of centuries ago.

The relationship of history and truth — as well as the connection between alchemy, chemistry and Paracelsian medical chemistry — is made clear if we refer once more to Bostocke's *Difference betwene the auncient Phisicke... and the latter Phisicke* published in 1585.[19] As a Paracelsian, Bostocke gave a definition of chemistry that is far removed from that employed today — and even his interpretation of alchemy had little in common with an art primarily devoted to the transmutation of the base metals. The opening sentence of his book makes this clear:

> The true and auncient phisicke which consisteth in the searching out of the secretes of Nature, whose study and use doth flowe out of the Fountaines of Nature, and is collected out of the Mathematicall and supernaturall precepts, the exercise whereof is Mechanicall, and to be accomplished with labor, is part of *Cabala,* and is called by auncient name *Ars sacra,* or *magna & sacra sciẽta,* or *Chemeia,* or *Alchimia, & mystica,* & by some of late, *Spagirica ars.*[20]

Thus for Bostocke true medicine may be equated with what we today would call science, and this in turn is nothing but mysticism in its fullest sense — Alchemy. In keeping with his age, and with alchemical tradition, he felt that the truth of Paracelsian medicine could be proven indisputable if it could be connected with the golden age of man. Bostocke presents us with a history of science that deplores innovation and instead views the achievements of his century as truly a Renaissance of the knowledge of the ancient seers. The history of science for Bostocke then was history with a purpose — to show that the alchemical-Paracelsian tradition is derived from an older source than the Galenic-Arabic medicine it was in competition with in the late sixteenth century. The significance he attached to his history may be gauged by the amount of space he devoted to the subject — ten chapters in a book that runs to a total of twenty-five chapters. [21]

Clearly these histories of chemistry and chemical medicine were important to their authors because they connected their science with Hermes and, occasionally, with Adam, thus providing a link with the divine knowledge given to man by God. How could the work of Aristotle or Galen compare with these credentials? However, with the decline of alchemy in the late seventeenth and the eighteenth centuries this reason for the histories was forgotten. For Boerhaave history was hardly necessary for a knowledge of the science, but he felt that nothing is «more interesting to an artist, than to know the rise and fate of his art.» [22] Throughout the eighteenth and nineteenth centuries a series of multi-volume histories of chemistry were written, many of them still of importance, such as J. F. Gmelin's three volume *Geschichte der Chemie* (1797-1799), [23] Hermann Kopp's four volume *Geschichte der Chemie* (1843-1847) which has been republished several times in this century, [24] and, more recently, E. O. von Lippmann's *Entstehung und Ausbreitung der Alchemie* (1919-1931). [25] At the same time important collections of texts were being prepared by M. Berthelot, [26] Wilhelm Ostwald [27] and the well known reprint series of the Alembic Club. [28] And as for bibliographical guides to the literature, there are few sciences that have comprehensive works to compare with John Ferguson's two volume *Bibliotheca Chemica* (1906), [29] H. C. Bolton's *Bibliography of Chemistry* (1893), [30] or the catalog of the Duveen collection, the *Bibliotheca alchemica et chemica*, [31] published in 1949.

Perhaps the last great comprehensive work in this tradition was *A History of Chemistry* written by James Riddick Partington (1886-1965). A physical chemist of considerable stature (he was appointed to the Chair of Chemistry at Queen Mary College at the age of thirty-three), Partington acquired an early interest in the history of chemistry. His first major work in the field was his *Origins and Development of Applied Chemistry* (1935) which contained over 25,000 references. [32] This was to be followed by his frequently reprinted *Short History of Chemistry* (1937) [33] and *A History of Greek Fire and Gunpowder* (1960). [34] The second, third, and fourth volumes of his monumental *A History of Chemistry* covering the sixteenth through the early twentieth centuries appeared between 1961 and 1964 — while the first part of the incomplete first volume was published posthumously in 1970. [35] This work alone totals some 3,000 pages and it will surely remain a major source of reference for both chemists and historians.

For Partington there seemed to be little need to argue for the value of the history of this science. Only in his Presidential Address to the British Society for the History of Science in 1951 did he digress to note that although Richard

Willstätter had emphasized the value of teaching chemistry on an historical basis, this attitude seemed to run counter to more recent trends.

> Some of us may forget at times that there is in existence a deep hostily to the study of that subject. We are made aware of this in many ways. The hostility is noticeable among some teachers in schools, who dislike books which touch upon the historical aspects of their subjects. In the universities we find that the subject of the history of chemistry has disappeared from the syllabus for the degree. When this happened, we were told that the time had come when chemistry must be treated on didactic lines, that the growth and complexity of the science were such that all the energies of students were absorbed in mastering the present state of the science, and that any mention of its origins was not only a waste of time but also could only confuse and repel the student approaching the subject.[36]

Needless to say, Partington disagreed with this development. His texts in inorganic chemistry included strong historical components and he clearly believed that this was the proper way to teach the subject. However, Partington had less interest in the history of science as an independent subject, and he surely had little concern with the relation of social factors to the development of chemistry. In his large *History* he discussed the scientific works of each author emphasizing chemical reactions and theories rather than the relation of these discoveries to broader scientific history. His work, impressive and important as it is, stands as a monument to the older internalist history.

If the history of chemistry has had such a long tradition of scientist-historians we must ask why it is that it has played such a small role in the development of the history of science as a whole. In an article on «The Historiography of Chemistry» in the December, 1983 issue of *History of Science* J. R. R. Christie and J. V. Golinski wrote that

> Chemistry in the seventeenth and eighteenth centuries retains a subsidiary position in the historiographical hierarchy. Worthy no doubt, but dull, she is courted by a few historians when they are faced with the competing enticements offered by her glamorously established sister, physics, or her exciting sisters, biological evolutionism (busily originating), and geology (busily rising). Chemistry attained the age of reason, quietly went about her business, and only latterly and briefly caused gossip by her liaison with a Frenchman, which made her mature.[37]

In short, tradition directed the history of chemistry to the chemist rather than to the newly emerging field of the history of science. Physics was the center of this discipline in the post-war years. The histories of geology and evolution have developed more recently and therefore in relationship to the newly established field. The history of chemistry, however, was a much older area of study, and one which had traditionally been directed toward chemists rather than to historians. It suffered further from its low ranking in Sarton's hierarchy of sciences.

Among historians of science with a special interest in the history of chemistry thirty years ago Douglas McKie stands out with special prominence. Professor of the history of science at University College, London, McKie devoted much of his research to Lavoisier and the Chemical Revolution of the eighteenth century.[38] He trained a group of students interested in similar problems and he did much to make this the reigning area of research in the history of chemistry. In France Daumas and in the United States Guerlac added to the strength of this area.[39] Guerlac's student, Marie Boas sought to introduce the history of chemistry as a fundamental subject for the historian of science in her *Robert Boyle and Seventeenth-Century Chemistry* (1958) in which she contrasted the earlier mystical views of the alchemists with Boyle's «rational mechanical theory in chemistry» in which a real attempt was made to explain chemical reactions in terms of the size, shape and motion of the small particles of the bodies. While not denying an eighteenth-century chemical revolution, Boas called for the recognition of a seventeenth-century chemical revolution as well because of the break between the work of Boyle and his colleagues with the alchemical and Paracelsian tradition.[40]

Emphasizing the modern elements in Boyle's thought Marie Boas titled the chapter on the mystical natural philosophies of the Renaissance in her *The Scientific Renaissance 1450-1630* (1962), «Ravished by Magic.»[41] And in a paper with her husband, A. R. Hall, on the alchemical manuscripts of Isaac Newton she argued that these were really chemical works — in the modern sense of the word — which had to be expressed in a traditional mystical language because no better language existed at that date.[42] Even Herbert Butterfield who had rebelled against positivistic historians in his Whig Interpretation of History (1931) wrote in *The Origins of Modern Science 1300-1800* (1949) that

> ...twentieth-century commentators on Van Helmont are fabulous creatures themselves, and the strangest things in Bacon seem rationalistic and modern in comparison. Concerning alchemy it is more difficult to discover the actual state of things, in that the historians who specialise in this field seem sometimes to be under the wrath of God themselves; for, like those who write on the Bacon-Shakespeare controversy or on Spanish politics, they seem to become tinctured with the kind of lunacy they set out to describe.[43]

That the appraisals of Boas and Butterfield were eventually rejected has been due primarily to Walter Pagel. Based upon nearly thirty years of research, his *Paracelsus: An Introduction to Philosophical Medicine in the Era of the Renaissance* appeared in 1958, the same year as Boas' book on Boyle.[44] However, rather than selecting only «modern» material from extensive writings as had some earlier scholars, Pagel sought to understand the «total man.» The end result was not the picture of a barbarous Renaissance magician, but rather, an influential figure who borrowed from many areas of science, medicine and philosophy then current. Showing clearly the influence of hermetic, neo-Platonic and Gnostic sources on the work of Paracelsus, Pagel pointed out that these philosophical themes which today seem anti-scientific, at that time stimulated men to a new observational approach to nature. Relying both on the special place of man in the Creation and the inter-related macrocosm and microcosm, Paracelsus insisted that man not only *could* study nature on his own — it was his pious *duty* to do this to learn of the goodness of his Creator. Throughout the Paracelsian literature we meet constant appeals for scholars to discard the works of the ancients and substitute for them fresh observations and personal experience.

This was all to have a considerable impact on the history of chemistry as well as the history of science. It became clear from Pagel's work that the Paracelsian cosmology and medicine was wedded to alchemical and chemical theory. Paracelsus described a divine Creation which was explained in terms of chemical separation and led to a new system of elementary principles, the *tria prima* or salt, sulphur and mercury. Geocosmic events were discussed in terms of chemical analogies and human physiology became a chemical physiology. Diseases were chemical malfunctions of the body and they were to be cured by chemically prepared medicines. The followers of Paracelsus demanded the downfall of the ancient authorities whose works were to be replaced with their own «philosophia nova» or «Chemical Philosophy.»

To be sure, the actual chemistry to be found in the work of Paracelsus is seldom of a high order. It seldom surpasses the work of the earlier alchemists, and even in the works of his followers chemical preparations are normally presented only for the sake of their pharmaceutical use. The primary goal of the Paracelsians was always a reformed medicine. However, it was also what they said it was, a Chemical Philosophy of Nature. As such it rivalled the more mechanical philosophies for roughly one hundred years. I will not dwell on this now for it will be the subject of my next lecture. However, I will add here that the work of Walter Pagel was not immediately accepted by historians of chemistry. This is due in large measure to the fact that his interest in Paracelsus was directed primarily to his medical philosophy. Consequently, he did not emphasize the Chemical Philosophy that forms such an important part of the Paracelsian corpus.

Aside from the far reaching implications of the Chemical Philosophy, historians of chemistry have newly reassessed the emergence of chemistry as an independent discipline. This problem is somewhat different from the establishment of the chemical world view of the Paracelsians. Owen Hannaway tacked this problem in *The Chemists and the Word* (1975) in which he examined the contrast between the Paracelsian apologist Oswald Crollius (1609) and Andreas Libavius (1597-1615). [45] The former sought to explain the chemical cosmology of the Paracelsians as a basis for the chemical medicine he advocated. The latter opposed the religious mysticism of Paracelsus and the alchemists even though he believed fervently in the possibility of transmutation. Rather, he sought to extract the chemical knowledge of earlier texts and to define and organize this material in a practical fashion for didactic purposes. The work of Libavius was indeed utilized in turn by Jean Beguin (1610) [46] and an important series of seventeenth-century teachers of chemistry many of whom prepared their own chemical textbooks which for the most part were to prove devoid of the mystical theory of the Paracelsian cosmologists.

More recently Christie and Golinski have reviewed the literature relating to seventeenth- and eighteenth-century chemistry. In effect they have rejected the traditional internalist approach that concentrated on the overthrow of the phlogiston theory and the recognition of the importance of the discovery of oxygen by Lavoisier. Rather, they lean favorably to the position of Hannaway that the emphasis on careful definitions and descriptions by Libavius and his teaching disciples must be further examined. However, we still know very little about the actual teaching of chemistry in the course of the seventeenth century. Christie and Golinski give several examples of significant changes that did occur relating to the organization of chemical theory and they emphasize the need to explore this factor more thoroughly. They close by noting that recent studies indicate that the Chemical Revolution of the eighteenth century is far more complex than older studies suggest. It «now appears to have been more to do with theories of the aeriform state and acidity and new concepts of composition.» [47] But no less important is the fact that like Beguin and the seventeenth century text book authors, Lavoisier wrote his own text book and collaborated on a new nomenclature. «As opponents... were not slow to realize, in learning the new words the student was learning equally, though less consciously, to reject the old concepts and theories» [48] I think that the message here is clear: if we are to learn more of the changes that have occurred in the history of chemistry, or the history of science for that matter, we must devote more time to textbooks employed and methods of teaching.

Let me add by way of a digression that José Maria López Piñero has placed great emphasis on the importance of the introduction of chemical medicine and its relationship to the new science in Spain. He admits that

> Estamos todavía muy lejos de disponer de um modelo sólidamente fundamentado de la integración de la actividad científica en la sociedad española de la época y de su trayectoria en relacíon con la desarrollada en el resto de Europa.

Nevertheless, he has shown the importance of the work of the Spanish Helmotian Juan de Cabriada and sees in the polemical debate between chemical physicians and Galenists the cause of the founding of the Sociedad de Medicina y Otras Ciencias de Sevilla in 1697.[50] Indeed, although López Piñero has noted the publication of a Spanish Paracelsian text as early as 1589, it would seem that the debate between the medical traditionalists and the innovators occurred in Spain roughly a century after it had reached its zenith in England, France, and Germany. This being the case, it would be a matter of considerable interest to follow the emergence of modern chemistry and its connections with a reformed medicine in the Iberian peninsula. It is only now that we are beginning to see the reprinting of significant texts for such a study. For instance, López Piñero has published an edition of Laurentius Coçar's *Dialogus Medicinae Fontes Indicans* (1589)[51] with notes and an introduction while the very early Spanish commentary on Lavoisier's *Nomenclature* written by Don Juan Manuel Arejula in 1788 has been republished with a detailed discussion by Ramón Gago and Juan Carrillo.[52]

Let me return to the work of Hannaway and the article by Christie and Golinski. The work of the former centers on the problem of the establishment of chemistry as a science independent of alchemy and Paracelsism. The approach suggested by Christie and Golinski suggests that different questions put to the texts of the seventeenth and the eighteenth centuries may well lead to a new interpretation of the growth of chemistry during that period. However, in both cases the assumption is made that the later alchemical and Paracelso-Helmontian texts need not be considered. If our interest should be primarily the establishment of chemstry as we know it they are right. However, there is no doubt that many alchemical and Paracelsian texts continued to be published throughout the eighteenth century. The author of the first table of affinity, Étienne François Geoffroy, read a paper against the increasing numbers of alchemists at the Parisian Academy of Sciences in 1722 while the Professor of Chemistry at Montpellier, Gabriel-François Venel, called for a new Paracelsus to rejuvenate chemistry in his article on that science in Diderot's *Encyclopédie.*[53]

It is at the end of the eighteenth century that we see not only the establishment of Lavoisier's new chemistry, but also a Romantic reaction against the severe mechanistic science that characterized much of the Enlightenment in Europe. This is a period when we witness a revived interest in alchemy and natural magic. It is also a period when we see Mesmer accused of plagiarizing the work of Paracelsus[54] and when we witness Hahnemann borrowing an important element of Paracelsian medical theory for his Homoeopathic Medicine.[55] For this reason I would argue that even though the Chemical Philosophy of the Renaissance did not form part of the mainstream chemistry of the eighteenth century, alchemical and Paracelsian books continued to find a ready public at that time. Furthermore, this literature seems to form an important connecting link with the rise of Naturphilosophie in the early nineteenth century.

Indeed, in many ways the study of nineteenth and early twentieth-century chemistry is in its infancy. To be sure, we have many older studies of importance including the fourth volume of Partington's *History of Chemistry* which presents us with nearly one thousand pages of technical data covering the period, but more recently there has been a turn toward the study of nineteenth-century chemistry in terms of intellectual history. David Knight has noted that

> Among [Sir Humphry] Davy's great achievements in the early years of the nineteenth century, in his own eyes and in those of his contemporaries, was to show that mechanics was not enough. Chemistry was a science of immaterial powers such as electricity as well as of matter.... Matter, on this analysis, was brutish and inert, and there was no real reason to suppose that there were different kinds of it. [56]

Here was the suggestion of a universal matter similar to much older theories. As he follows his subject, Knight pictures «a surprisingly close and shifting contact between chemical theories and philosophical positions.» A central question regarded that of the existence of atoms, and the early twentieth century work of Thomson, Curie and Rutherford

> seemed to confirm what many of the greatest chemists of the nineteenth century had believed; that our simple and harmonious world was not constructed of atoms of numerous different kinds, but simply of particles of matter all of the same kind. [57]

In Germany Reinhard Löw has embarked upon a study of the role of chemistry in German Naturphilosophie. [58] He has documented the fact that the development of organic chemistry was in no way hindered in its progress by this intellectual movement. Thus, although we are all familiar with Liebig's attack on the mysticism of the Naturphilosophen, it would seem that this may have improperly diverted us from the study of this movement. I think that we can safely predict that Naturphilosophie will become an area of future concern not only for historians of science, but very specifically for historians of chemistry.

Research on nineteenth-century chemistry is also beginning to reflect more recent contextual trends. The rise of chemical societies has been a matter of interest to those concerned with the professionalization of the field. Others have been interested in the rapid development of new apparatus and techniques. The rise of the chemical laboratory beginning with Liebig has been essential to the development of modern chemical education which grew rapidly in the course of the century because of practical needs. In the United States in particular the new land grant colleges always emphasized chemistry because of the need to train agricultural chemists who could analyze soil samples. [59]

The rise of chemical industry is no less important for the historian. Here special interest has centered on organic chemistry and the rise of the dye industry.[60] Although the first chemically prepared dye, mauve, was produced by Perkin in England, it was not to be developed commercially by English manufacturers. Instead, this knowledge was carried to Germany by Perkin's students where it became the basis of the German chemical industry in the late nineteenth century. This industry was to prove of great value to Germany at the outbreak of the first World War. In particular, the discovery of the synthesis of ammonia by Fritz Haber in 1908 was rapidly developed to the production stage and this was to free Germany from foreign supplies of nitrates in the production of munitions.[61] Haber himself was to work for the government on poison gases and their use in attack procedures during the war. The agency which he headed was eventually to include one hundred and fifty university personnel and some two thousand assistants.

Of course, during peace the Haber process was to benefit farmers everywhere through the development of soil additives and Haber himself received the Nobel Prize, but this isolated case does present us with an historiographical problem. In traditional fashion, J. R. Partington discussed the Haber process, but only as an important chemical discovery.[62] Its uses were of little concern to him since the uses of the process were outside of his concept of the history of the chemistry. Once again we find ourselves back to the internal-external concerning the history of science.

Is it possible for us to say that external factors can affect the growth of science? Surely in the case of commerce as in the example of the dye industry it is possible to train large numbers of scientists to develop a promising area of research. We know also that the German chemical industry in the late nineteenth century did this not only through the establishment of industrial laboratories, but also through the donation of money for university research in scientific fields in which industry had a special interest. In the case of the first World War we see the immediate recognition of the importance of the Haber process. Recent research has further established the fact that the Weimar Republic established a scientific agency that began funding research grants considered to be of value as early as the 1920s.[63] There is little doubt then that external factors can concentrate research in certain promising fields.

But this is not a twentieth-century development in itself. Francis Bacon envisaged a «New Philosophy» that would result in the improvement of practical processes — while the French Academy of Sciences had as one of its objectives the utilization of scientific expertise for the benefit of the state. Also in the seventeenth century Johann Rudolph Glauber forecast economic prosperity for farmers to convert their excess wheat and wine in times of plenty to concentrates that could be reconstituted to beer, bread and wine in times of want. He argued further for the development of chemical methods to obtain minerals from Germany's vast forests as well as a concentrated effort to establish a commercially viable method for the production of gold from the base metals. Above all, he wrote of the need for chemical weapons such as acid mists which might be used to blind enemy soldiers. This seemed so important that he urged the establishment of a chemical academy that would constantly seek to improve existing weapons and invent new ones. Only in this fashion would it be possible to keep technologically ahead of one's enemies.[64]

There is little doubt then that external factors involving politics, warfare, commerce, economics and religion — among others — play an important role in relating science to society. This can be shown over a very long period. Furthermore, it is possible to channel research into promising fields by the means of financial aid and this may well accelerate scientific advance. However, it has yet to be proven that such external factors affect scientific discovery or the development of major theories.

Let me conclude this lecture by noting that the writing of the history of chemistry has undergone a revolution of its own in the past quarter century. Originally the history of this science was written primarily for the benefit of chemists. In this form it has a long history which may be traced back to the alchemists and the Paracelsian medical chemists who used such histories to connect their knowledge with the pristine wisdom given to man by God. Such histories were used to show the greater antiquity — and therefore the greater authority — of the Chemical Philosophy over that of the Aristotelians and Galenists. In modern form — devoid of its original polemical purpose — this tradition is seen best in the monumental work of Partington, a vast compendium of historical facts about chemistry which are for the most part unrelated to other intellectual or social currents.

If I had to choose one date as the turning point in the historiography of chemistry I would probably choose 1958 since that year saw the publication of Marie Boas' first book on Robert Boyle and Walter Pagel's book on Paracelsus. The first of these sought to bring seventeenth-century chemistry into the more general picture of the Scientific Revolution by showing that Boyle was not only a chemist, but that he was also one of the most influential mechanical philosophers. In a sense one might say that she made chemistry worthy of study by making it a part of physics. Pagel's book was different. He appraised the total work of Paracelsus and placed his medical philosophy in the intellectual context of the period. More important for us, he showed the deep impact of chemistry on his medical and cosmological thought. Others — including myself — have since expanded on this theme to show the existence of a Chemical Philosophy of nature that many thought would become the «new science» of the Scientific Revolution. In short, the work of Boas and Pagel — each in its own way — demanded the integration of chemistry into the Scientific Revolution as a whole.

The question of the separation of chemistry as a distinct discipline from alchemy and the Chemical Philosophy of the Paracelsians has proved to be a different problem. Seeking an answer to this, Hannaway examined the contrast between the Paracelsian, Crollius, and his chemist opponent, Andreas Libavius. The latter sought to strip the mysticism from chemistry and to replace it with clear definitions and descriptions of chemical processes. Although the chemical cosmology remained influential throughout the seventeenth century, the work of Libavius led to well known chemical textbooks and to a teaching tradition that presented the subject in an understandable fashion. More recently Christie and Golinski have noted the need to further document the chemical teaching tradition of the seventeenth and eighteenth centuries, suggesting that when this is done we may arrive at a different interpretation of Lavoisier and of the Chemical Revolution itself.

I believe then that it is true to say that recent research is transforming our understanding of the history of chemistry. Unfortunately, the research on nineteenth and twentieth century chemistry is not yet nearly as advanced. Here we can only point to areas of research in which preliminary studies promise future work of great significance. Some of these relate to internalist studies. I am thinking now of recent papers on the origins of bio-chemistry and detailed studies of physical chemistry.[65] However, there are other recent trends that relate this science to a much broader historical context. I have been interested myself in a persistent interest in alchemy that continued throughout the Enlightenment and seems to connect with the Romantic movement and Naturphilosophie. Others have related nineteenth-century chemistry to other aspects of intellectual history. There is also an increasing interest in the professionalization of chemistry, a subject that includes the establishment of chemical societies and the development of chemical education. And it is a subject of great importance for history as a whole to take into account the rise of chemical industry in the nineteenth century and to show the relationship of this science to the state.

In short, although histories of chemistry have been written for at least four hundred years, there is still much to be done. It is true that we now have many of the facts needed to rewrite our internalist histories, but we still know all too little about the role of chemistry and its technology in terms of intellectual, political, and social history. I know now that we cannot understand the Scientific Revolution without an understanding of the role played by chemistry. But I believe in the future when we know more than we do now that we will also be able to say that it is impossible to understand recent world history without a knowledge of the history of chemistry.

NOTES

1. At Northwestern University I studied the history of chemistry under Frank T. Gucker, and at Indiana University I studied it under Gerald Schmidt. Both gave courses on the subject through the Department of Chemistry.

2. J. R. Partington, *A Text-Book of Inorganic Chemistry* 6th ed. (London: Macmillan, 1950).

3. M. Cannon Sneed and J. Lewis Maynard, *General Inorganic Chemistry,* 5th printing (New York: Van Nostrand, 1945).

4. H. E. Roscoe and C. Schorlemmer, *A Treatise on Chemistry,* (4th ed. 2 vols., 1911, 1913), *1,* pp. IX-XV, «Historical Introduction.»

5. Berthelot's historical studies are so numerous and important that they hardly need be listed here. Similarly Hermann Kopp's Geschichte der Chemie (4 vols., 1843-1847; reprinted Hildescheim: Olms, 1966) remains indispensable to the historian of chemistry. Only slightly less well-known than these monumental works are Ramsay's *Gases of the Atmosphere,* 4th ed. (London, 1915) and Schorlemmer's *Rise and Development of Organic Chemistry* 2nd ed. (London: Macmillan, 1894).

6. William Henry, *The Elements of Experimental Chemistry* (1st American from the 8th London edition) 2 vols., (Philadelphia: Robert Desilver, 1819) *1,* p. IX.

7. Douglas McKie, «On Some MS. Copies of Black's Chemical Lectures — III,» *Annals of Science 16* (1960), 1-9 (1).

8. Olaus Borrichius, *De ortu et progressu chemiae dissertatio* as printed in the *Bibliotheca chemica curiosa,* ed. Jean Manget, 2 vols. (Geneva: Chouet, G. De Tournes, et al., 1702), *1,* pp. 1-37 (12).

9. Hermann Conringius, *De Hermetica Medicina Libri Duo...* 2nd ed. (Helmstedt: typis & sumptibus H. Müller, 1669), p. 345.

10. See the review of Olaus Borrichius, *Hermetis, Aegyptiorum, et chemicorum sapientia ab Hermanni Conringii animadversionibus vindicata per Olaum Borrichium* (Hafniae: sumptibus P. Hanboldi, 1674) in the *Philosophical Transactions of the Royal Society of London 10* (n.° 113) (April 26, 1675), 296-301 (297).

11. In addition to the reference in note 10 see the review of the *De ortu et progessu chemiae* in the *Philosophical Transactions 2* (n.° 39) (1668), 779 and the *Journal des Sçavans 3* (1675), 209-11.

12. The exchange was referred to as important by Hermann Boerhaave in his introductory lecture on the history of chemistry while two historical works by Borrichius (including the De ortu et progressu chemiae) began Manget's fundamental alchemical collection, the *Bibliotheca chemica curiosa.*

13. The texts of Beguin and Lemery are not historically oriented, but Lefèvre, *(A Compleat Body of Chymistry* [London: O. Pulleyn Jr. to be sold by John Wright, 1670]), pp. 1-4 gives a short history in his preface.

14. Petrus Severinus, *Idea Medicinae Philosophicae* (Hagae-Comitis: Adrian Clacq, 1660), pp. 1-5.

15. Daniel Sennert, *De Chymicorum cum Aristotelicis et Galenicis Consensu ac Dissensu* 3rd ed. (Paris: Apud Societatem, 1633), pp. 17-28.

16. Robertus Vallensis, «De veritate & antiquitate artis Chemicae & pulveris sive Medicinae Philosophorum vel auri potabilis, testimonia & theoremata ex variis auctoribus,» *Theatrum Chemicum,* et. L. Zetzner, 6 vols. Strassburg: Zetzner, 1659-1661), *1,* pp. 7-29.

17. See above, note 8

18. McKie, *op. cit.,* pp. 1ff.

19. For a general discussion of Bostocke's work see Allen G. Debus, «The Paracelsian Compromise in Elizabethan England,» *Ambix 8* (1960), 71-97, especially pp. 77-84. Bostocke's views on the prime matter have been examined by Walter Pagel in «The Prime Matter of Paracelsus,» *Ambix 9* (1961), 117-135 (124-129).

20. R. B. (Bostocke), Esq., *The difference betwene the auncient Phisicke... and the latter Phisicke* (London: Robert Walley, 1585), sig. B1^{r}.

21. The historical chapters have been reprinted with an introduction and notes by Allen G. Debus, «An Elizabethan History of Medical Chemistry,» *Annals of Science, 18* (1962), 1-29.

22. Hermann Boerhaave, *A New Method of Chemistry; Including the Theory and Practice of that Art: Laid down on Mechanical Principles, and accommodated to the Uses of Life. The whole making a Clear and Rational System of Chemical Philosophy,* trans. P. Shaw and E. Chambers (London: J. Osborn and T. Longman, 1727), p. v.

23. Johann Friedrich Gmelin, *Geschichte der Chemie seit dem Wiederaufleben des Wissenshaften bis an ende des achtzehneten Jahrhunderts,* 3 vols. (Göttingen: J. G. Rosenbusch, 1797-1799).

24. See above, note 5.

25. E. O. von Lippmann, *Entstehung und Ausbreitung der Alchemie,* 2 vols. (Berlin, 1919, 1931). Von Lippmann also authored volumes on the history of organic chemistry and the history of science in general.

26. See for instance his three volume *Collection des anciens alchimistes Grecs* (1888; reprinted London: The Holland Press, 1963), a work that is only today being replaced by the research of Robert Halleux [*Les Alchimistes Grecs* (Paris: Société d'Édition «Les Belles Lettres»)]. Here the first volume (containing the Leyden Papyrus, the Stockholm Papyrus and various fragments) appeared in 1981]. See also Berthelot's three volume *La chimie au moyen age (1893; reprinted Osnabruck: Otto Zeller and Amsterdam: Philo Press, 1967).*

27. *Wilhelm Ostwald, ed. Klassiker der exacten Wissenschaften,* 7 vols. (Leipzig, various dates).

28. *Alembic Club Reprints,* 22 vols. (Edinburgh: E. and S. Livingstone, various dates).

29. J. Ferguson, *Bibliotheca chemica,* 2 vols. (Glasgow, 1906; London: Academic and Bibliographical Publications, 1954).

30. Henry Carrington Bolton, *A Select Bibliography of Chemistry 1492-1892* Smithsonian Miscellaneous Collections, 850, 1893; reprinted New York: Kraus Reprint Corporation, 1966).

31. Denis I. Duveen, *Bibliotheca alchemica et chemica* (London: Dawsons, 1949).

32. J. R. Partington, *Origins and Development of Applied Chemistry* (London: Longmans, Green & Co., 1935).

33. J. R. Partington, *A Short History of Chemistry* (1937; 2nd edition, London: Macmillan, 1951).

34. J. R. Partington, *A History of Greek Fire and Gunpowder* (New York: Barnes and Noble, 1960).

35. J. R. Partington, *A History of Chemistry* [vol. 1, Part 1: Theoretical Background (London: Macmillan; New York: St. Martin's Press, 1970); vol. 2: 1500-1700 (1961); vol. 3: 1700-1800 (1962); vol. 4: 1800 to the present time (1964)].

36. J. R. Partington, «Chemistry as Rationalised Alchemy» (Presidential Address to the British Society for the History of Science delivered by the late Professor J. R. Partington, M.B.E. on 7th May 1951) in *A History of Chemistry,* vol. 1, Part 1, pp. XI-XVIII (XV).

37. J. R. R. Christie and J. V. Golinski, «The Spreading of the Word: New Directions in the Historiography of Chemistry 1600-1800,» *History of Science 20* (1982), 235-66 (235).

38. See for example his two books on Lavoisier, *Antoine Lavoisier: The Father of Modern Chemistry* (Philadelphia: J. B. Lippincott, n.d. [1935]) and *Antoine Lavoisier: Scientist, Economist, Social Reformer* (New York: Henry Schuman, 1952).

39. Maurice Daumas, *Lavoisier, théoricien et expérimentateur* (Paris, 1955) and numerous articles, the most important of which are cited in the bibliography to Henry Guerlac, *Antoine-Laurent Lavoisier: Chemist and Revolutionary* (New York: Charles Scribner's Sons, 1975). See also Guerlac's *Lavoisier — The Crucial Year: The Background and Origin of His First Experiments on Combustion in 1772* (Ithaca: Cornell U.P., 1961). For a collection of Guerlac's papers see *Essays and Papers in the History of Modern Science* (Baltimore and London: The Johns Hopkins U.P., 1977).

40. Marie Boas (Hall), *Robert Boyle and Seventeenth-Century Chemistry* (Cambridge: Cambridge U.P., 1958), pp. 231-32.

41. Marie Boas, *The Scientific Renaissance 1450-1630* (New York: Harper, 1962), pp. 166-96.

42. Marie Boas, and A. R. Hall, «Newton's Chemical Experiments,» *Archives internationales d'histoire des sciences 11* (1958), 113-52.

43. Herbert Butterfield, *The Origins of Modern Science 1300-1800* (New York: Macmillan, 1952), p. 98.

44. Walter Pagel, *Paracelsus: An Introduction to Philosophical Medicine in the Era of the Renaissance* (Basel: Karger, 1958). Pagel's influence may be seen clearly in the breadth of the papers published in *Science, Medicine and Society in the Renaissance: Essays to Honor Walter Pagel* ed. Allen G. Debus, 2 vols. (New York: Science History Publications, 1972).

45. Owen Hannaway, *The Chemists and the Word: The Didactic Origins of Chemistry* (Baltimore and London: The Johns Hopkins U.P., 1975).

46. Andrew Kent and Owen Hannaway, «Some New Considerations on Beguin and Libavius,» *Annals of Science 16* (1960, published 1963), 241-51.

47. Christie and Golinski, *op. cit.,*259.

48. *Ibid.*

49. José María López Piñero, *Ciencia y Técnica en la Sociedad Española de los Siglos XVI y XVII* (Barcelona: Editorial Labor, 1979), p. 12.

50. José María López Piñero, *La Introduccíon de la Ciencia Moderna en España* (Barcelona: Ediciones Ariel, 1969), pp. 108-17.

51. José María López Piñero, *Le «Dialogus» (1589) del Paracelsista Llorenç Coçar y la Cátedra de Medicamentos Químicos de la Universidad de Valencia (1591)* (Valencia: Cátedra e Instituto de Historia de la Medicina, 1977).

52. Ramon Gago and Juan L. Carrillo, *La Introduccion de la Nueva Nomenclatura Quimica y el Rechazo de la Teoria de la Acidez de Lavoisier en España: Edición facsímil de las Reflexiones sobre la nueva nomenclatura química (Madrid, 1788) de Juan Manuel de Aréjula* Malaga: Universidad de Malaga, 1979).

53. Allen G. Debus, «The Paracelsians in Eighteenth-Century France: A Renaissance Tradition in the Age of the Enlightenment,» *Ambix 28* (1981), 36-54; Allen G. Debus, «Alchemy and Paracelsism in Early Eighteenth- Century France,» to be published in Ingred Merkel and Allen G. Debus eds., *Hermeticism and the Renaissance: Papers Presented at the International Conference at the Folger Library 25-27 March 1982.*

54. Allen G. Debus, «History with a Purpose: The Fate of Paracelsus,» (Inaugural Lecture for the International Congress for the History of Pharmacy, Washington, D.C., 22 September 1983).

55. The belief that «like cures like,» fundamental to Hahnemann, was also one of the major dictums of Paracelsus opposed to Galenic medicine. While Hahnemann did not acknowledge the Paracelsian tradition, early historians of homoeopathic medicine were aware of some pre-Hahnemann sources. See the list of earlier references by A. Gerald Hull appended to Rev. Thomas R. Everest, *A Popular View of Homoeopathy...* (New York: William Radde, 1842), pp. 233-36.

56. David M. Knight, *The Transcendental Part of Chemistry* (Folkestone: Dawson, 1978), p. vi.

57. *Ibid.,* pp. i-ii.

58. Reihnard Löw, «The Progress of Organic Chemistry During the Period of German Romantic Naturphilosophie (1795-1825),» *Ambix 27* (1980), 1-10; *Pflanzen-chemie zwischen Lavoisier und Liebig* (München: Straubing, 1977).

59. For example see Roger Hahn, *The Anatomy of a Scientific Institution: The Paris Academy of Sciences, 1666-1803* (Berkeley/Los Angeles/London: University of California Press, 1971); Jack Morrell and Arnold Thackray, *Gentlemen of Science: Early Years of the British Association for the Advancement of Science* (Oxford: Clarendon Press, 1981); J. C. Cutter, «The London Institution (1805-1933),» Ph.D. dissertation, University of Leicester, Supervisor, W. H. Brock; Margaret W. Rossiter, *The Emergence of Agricultural Science: Justus Liebig and the Americans 1840-1880* (New Haven: Yale U.P., 1975).

60. John J. Beer, *The Emergence of the German Dye Industry* (Urbana: University of Illinois Press, 1959).

61. K. F. Bonhoeffer, «Fritz Haber» in *Great Chemists, Eduard Farber, ed. (New York/London: Interscience, 1961), pp. 1299-1312 (1305).*

62. *Partington, History of Chemistry 4,* p. 636.

63. Paul Forman, «Scientific Internationalism and the Weimar Physicists: The Ideology and Its Manipulation in Germany after World War I,» *Isis 64 (1973), 151-80; «Weimar Culture, Causality, and Quantum Theory 1918-1972: Adaptation by German Physicists to a Hostile Intellectual Environment,» Historical Studies in the Physical Sciences 3* (1971), 1-115.

64. Allen G. Debus, *The Chemical Philosophy: Paracelsian Science and Medicine in the Sixteenth and the Seventeenth Centuries,* 2 vols. (New York: Science History Publications, 1977), *2,* pp. 425-41.

65. An excellent example is to be found in the recently published *From Medical Chemistry to Biochemistry: The Making of a Biomedical Discipline* (Cambridge/London/New York: Cambridge U.P., 1982) by Robert E. Kohler.

LECTURE 4

THE SCIENTIFIC REVOLUTION: A CHEMIST'S REAPPRAISAL[1]

I have tried to indicate in my earlier lectures that the traditional interpretation of the Scientific Revolution presents us with a relatively straightforward picture. In it the rise of modern science is most commonly pictured as a conflict of «Ancients» and «Moderns,» that is, those who remained faithful to Aristotle in Natural Philosophy and to Galen in Medicine against those who espoused a certain «New Philosophy» or a «New Science» founded on fresh observations and experiments. It is the latter group that is normally associated with the Mechanical Philosophy of the seventeenth century.

The main thread of this story leads us from one great figure to another, through the work of Copernicus, Tycho Brahe, Kepler, Galileo, and Sir Isaac Newton. Of course William Harvey is mentioned as well because of his discovery of the circulation of the blood, but, in general, biological developments are secondary to the restatement of Ptolemaic cosmological theory and the gradual solving of the new problems of the physics of motion associated with a moving earth. Bacon and Descartes are normally discussed also, but generally to show that neither of their scientific methodologies are satisfactory for science as we know it today. In short, the student is presented with a story of scientific progress that resulted in the establishment of the Copernican system and the foundation of classical mechanics.

If we hope only to follow the steps leading to the foundation of classical mechanics then we will see little problem with this traditional approach to the field. However if we seek to understand nature in terms of those scholars who lived in the period of the Scientific Revolution we may well find the traditional accounts unsatisfactory. If we try to understand their work from their point of view we will not be able to be as selective in our choice of subject matter since we will have to be led by their questions as well as our own. In short, we must decide whether we are to be «modernist» or «contextual» in our interpretation.

If we do turn to the broad spectrum of texts written at the time we soon become aware that the concerns of authors interested in a new philosophy of nature went far beyond astronomy and the physics of motion. I think that there is little doubt that the most hotly debated subjects relate to medicine and chemistry — perhaps not defined in our terms, but rather, in terms of a Chemical Philosophy of nature that was proposed as a proper replacement for the works of Aristotle and Galen still being taught at the universities. Even when we speak of «ancients» and «moderns» we must be careful for when Libavius speaks of the «Philosophia nova» he uses this term to characterize the Paracelsian whom he attacked. We must then be willing to examine other areas than physics if we wish to reevaluate the Scientific Revolution as a whole.

Of special interest for the historian of chemistry is the work of the Swiss-German physician, Paracelsus (1493-1541), and his followers. Their work is of interest to us today less for specific medical reforms than for their search for a new philosophy of nature based upon chemistry. Paracelsus had been influenced by traditional alchemy, by medical theory and practice, and by Central European mining techniques, but he went beyond all of these to develop an all encompassing view of nature. He did believe in transmutation, but for him this goal was of far less importance than medical alchemy. This meant the use of chemical methods for the preparation of drugs, but in addition it meant a mystical alchemical approach to medicine that might apply to macrocosmic as well as to microcosmic phenomena.

The Paracelsians of the sixteenth century differ from other nature philosophers of the period in their emphasis on the importance of medicine and alchemy as bases for a new understanding of the universe. Characteristic of the Paracelsians was their firm opposition to the dominant Aristotelian-Galenic tradition of the universities. They rejected logic as a guide to truth — and therefore mathematical abstraction,[2] they sought an alternative to the Aristotelian elements[3] — surely a mainstay of Scholastic natural philosophy, and they found no value in the humoral medicine of the Galenists.[4] Rather, in their firm rejection of the Scholastic tradition, they emphasized the religious nature of their quest for knowledge and they claimed hidden truths they thought had not been fully recognized earlier in the Hermetic and neo-Platonic texts of late antiquity. The Paracelsian was convinced that man must seek an understanding of his Creator through the two books of divine revelation: the Holy Scriptures and the Book of Creation or Nature.[5] The latter would be understood only through fresh and unprejudiced study in the field and in the laboratory rather than through a rehash of old books in disputations. Thus we find the early Paracelsian, Peter Severinus, telling his readers in 1571 to

> sell your lands, your houses, your clothes and your jewelry; burn up your books. On the other hand, buy yourselves stout shoes, travel to the mountains, search the valleys, the deserts, the shores of the sea, and the deepest depressions of the earth; note with care the distinctions between

> animals, the differences of plants, the various kinds of minerals, the properties and mode of origin of everything that exists. Be not ashamed to study diligently the astronomy and terrestrial philosophy of the peasantry. Lastly, purchase coal, build furnaces, watch and operate with the fire without wearying. In this way and no other, you will arrive at a knowledge of things and their properties.[6]

Indeed, the Paracelsians constantly called for a new observational approach to nature, and for them chemistry or alchemy seemed to be the best example of what this new science should be.

The Paracelsians were quick to offer an alchemical interpretation of *Genesis.*[7] Here they picture the Creation as the work of a divine alchemist separating the beings and objects of the earth and the heavens from the unformed *prima materia* much as the alchemist might distill pure quintessence from a grosser form of matter. On this basis it was possible to postulate the continued importance of chemistry as a key to nature.

The search for physical truth in the Biblical account of the Creation focused special attention on the formation of the elements.[8] Paracelsus regularly used the Aristotelian elements, but he also introduced the *tria prima* — the principles of Salt, Sulphur, and Mercury. The latter were a modification of the old sulphur-mercury theory of the metals employed by Islamic chemists, but they differed from the older concept in that they were to apply to all things than be limited to the metals alone. The introduction of these principles had the effect of calling into question the whole framework of ancient medicine and natural philosophy since these had been grounded upon the Aristotelian elements. Unfortunately, the fact that Paracelsus did not clearly define his principles tended to make this problem an ill-defined one permitting each chemist to make his own interpretation.

In their attempt to understand the universe through chemical observations or analogies, the Paracelsians followed an old tradition, but one that was secondary in the alchemical literature to both gold-making and the preparation of chemical medicines.[9] For the sixteenth-century Paracelsians it was commonplace to interpret both macrocosmic and microcosmic phenomena chemically. Thus, in the great world they explained meteorological events in terms of chemical analogies. On the geocosmic level they argued over differing chemical interpretations of the growth of minerals and the origin of mountain springs.[10] And in their search for agricultural improvements they postulated the importance of dissolved salts as the reason for the beneficial result of fertilizing with manure.[11] This theory proved to be the basis of later developments that were to result in a true agricultural chemistry.

It is not surprising that the Paracelsians should have approached their medicine as chemists. They felt assured that their knowledge of the macrocosm might be properly applied to the microcosm.[12] Thus, if an aerial sulphur and niter were the cause of thunder and lightning in the heavens, the same aerial effluvia might be inhaled and generate burning diseases in the body. Humoral medicine was rejected *in toto.* Rather than discuss an imbalance of the bodily fluids leading to disease as did the Galenists, these physicians spoke rather of disease forming substances entering the body through the air or through food — lodging in some organ and causing organic disfunction. This was explained most often in terms of internal archei — forces within individual organs operating much as live alchemists in their laboratories.[13] When they failed to operate properly it would be impossible to properly eliminate impurities from the body and disease would result.

The need for medical reform was underscored by the fact that the Renaissance was a period that witnessed new and violent diseases. The Chief of these were the venereal diseases. These chemical physicians stated that their new stronger remedies — often prepared chemically from metals — were essential for the proper cure of these truly terrifying illnesses. There is little doubt that the sixteenth century saw important changes in the chemical preparation of medicines.[14] Paracelsus himself still reflected medieval distillation techniques and his preparations are largely characterized by the search for a distilled quintessence — frequently this proved to be the original solvent. The true chemical products of the reaction were often discarded. But, by the turn of the century such practices were undergoing change. Iatrochemists now turned less to distilled quintessences and more to precipitates and residues in their search for new remedies. This was to be essential for the understanding of chemical reactions.

But interesting as the details of this Chemical approach to nature are, they would remain unimportant if they had been ignored at the time. In fact, in the century between 1550 and 1650 conflicts between the Paracelsians on the one hand and the Aristotelians and the Galenists on the other were common. We must establish this point if we are to argue for their inclusion in accounts of the Scientific Revolution. As the tracts of Paracelsus were gradually recovered and published in the middle decades of the sixteenth century they began to win converts to the new system. Soon systematized condensations of the rambling discourse of Paracelsus — such as that of Peter Severinus — were required for the chemical practitioners.

The growing numbers of chemical physicians made this school the subject of a lengthy attack by the learned Thomas Erastus who served both as Professor of Theology at Heidelberg and as Professor of Theology and Moral Philosophy at Basel.[15] In his *Disputationes de Medicina Nova Paracelsi* (1572-1574) he pictured Paracelsus primarily as an ignorant man driven by ambition and vanity, a *magus* informed by the devil and evil spirits. He condemned him as an untrustworthy charlatan who continually contradicted himself, a man who knew no logic and consequently wrote in a totally disorganized and incomprehensible manner.

Erastus was wholly opposed to the philosophic system of Paracelsus. It was impious to compare the divine Creation to a chemical separation or to interpret the universe in terms of the macrocosm and the microcosm. Hardly less objectionable were the three principles of Paracelsus which, Erastus asserted, were not at all elementary in nature. Fire analyses and distillation procedures had produced them through the action of heat and one could therefore rest assured that the Aristotelian elements remained the basic substances of nature.

Erastus objected also to the views of Paracelsus on disease. Here he argued that to assume that diseases are separate entities which enter man from without was unthinkable and he reaffirmed that the traditional humoral medicine was truly the crowning glory of the Galenists. Hardly less objectionable was the role of innovator assumed by Paracelsus in his chemically prepared medicines. His theory of cure had led to the use of all kinds of minerals and metallic substances — and he pointed particularly to the poisonous mercury compounds — which were nothing less than lethal poisons. How then was it possible that so many men were drawn to this medical heresy by the reports of the wonderful cures performed by Paracelsus? The fact is, Erastus answered, that these «cures» were at best temporary ones. In his search through the archives at Basel he was gratified to find that all those treated there with Paracelsian methods had died within a year even if they had shown an initial improvement.

In France a conflict between the Chemists and the Galenists resulted in a series of law suits regarding the use of metallic substances internally as medicines.[16] An initial victory by the Galenists in 1566 proved to be premature and this was followed by bitter dispute in Paris throughout the remainder of the century. At that time, a new series of books and pamphlets penned by Joseph Duchesne (1544-1609) in defense of the new chemicals resulted in further controversy on a subject which now inflamed authors from all parts of Europe.

Early in the new century Duchesne published two important books in which he defended the chemical remedies and the Paracelsian approach to nature. Pointedly he contrasted the Galenists with the Chemists. The former, he wrote, follow Galen «and as if by a royal decree where the sentence is form and without doubt,» they pronounce their medical decrees. The chemists, however, place their faith not in books, but in reason and experience. This is the real basis of chemical medicine and the root of the disagreement between the Galenists and the Paracelsians.

Although Duchesne did pay lip service on occasion to both Galen and Hippocrates, it was clear that he hoped to ground both natural philosophy and medicine on the study of chemistry. It is understandable then that his more conservative medical colleagues might well feel threatened. Such indeed was to be the case. Duchesne's books were immediately answered in a condemnation written by the elder Jean Riolan — and the remainder of the decade was to witness the publication of a large number of pamphlets, monographs, commentaries and volumes that alternately defended and attacked Galenic, Paracelsian and chemical medicine. By 1605 the dispute was widely known beyond the borders of France and a number of the polemical tracts had been translated into other languages. An English version of Duchesne's books appeared as early as 1605 while in Germany Andreas Libavius (1540-1616) published a lengthy refutation of the censure of the chemists by the Parisian school of medicine. The younger Riolan wrote a response to this work — and this brought forth an *Alchymia triumphans* in 1607 which was a sentence by sentence reply to the French physician in a volume over nine hundred pages in length.

Joseph Duchesne died in 1609, but the debate was far from over. Six years earlier the first to defend him publicly had been his friend, Theodore Turquet de Mayerne (1573-1655). For this action the medical faculty had forbidden his colleagues to consult with him while Henry IV was urged to rescind his public offices. Outcast by the medical establishment in Paris, this same Turquet de Mayerne was invited to settle in England as Chief Physician to King James I in 1610.

In London the situation was far different.[17] For several decades the College of Physicians had been discussing the publication of an official pharmacopoeia. Their purpose was perhaps less for the benefit of the medical profession than it was to exert control over the pharmacists, the distillers and the myriad unlicensed practitioners who preyed on the sick. But, regardless of their reasons, it is clear that from the beginning, the Fellows of the London College were willing to accept those chemicals that proved to be of value medically.

Mayerne was to thrive in his new home. Elected a Fellow of the College, he worked diligently for the official acceptance of chemicals in medicine. The result was the first official national pharmacopoeia which was issued in 1618. The author of the preface (almost surely Mayerne) pointed out that although they venerated the learning of the ancients and listed their remedies, «we neither reject nor spurn the new subsidiary medicines of the more recent chemists and we have conceded to them a place and corner in the rear so that they might be as a servant to the dogmatic medicine, and thus they might act as auxiliaries.». The *Pharmacopoeia Londinesis* was a compromise effort and relatively few of the listed chemicals were highly controversial in nature. Still, their inclusion in the first official national pharmacopoeia clearly gave them a status they had not previously enjoyed — and at the same time the College had asserted its control over the acceptance of specific chemical medicines and the determination of the approved methods of preparation.

If Turquet de Mayerne represents the practical chemist urging the adoption of the new medicines, Robert Fludd (1574-1637) represents the mystic.[18] Fludd, no less than Crollius, may be connected with Andreas Libavius. In 1614 and 1615 several anonymous pamphlets ascribed to an otherwise unidentified Rosicrucian Brotherhood presented a call for religious and educational reform.[19] The educational reforms were Paracelsian in tone and oriented toward medicine and chemistry. They were quickly translated into numerous languages and they elicited an astonishing response. Among those who attacked them was Libavius who favored the new chemical medicines, but saw in these works an adherence to the same sort of Paracelsian mystical cosmology that he opposed in the work of Oswald Crollius.

Fludd, on the other hand, hoped for the acceptance of a mystical world view based on a fusion of Hermetic and Christian doctrine. He was delighted by the Rosicrucian tracts and he sought to contact the Brotherhood by writing a sharp reply to Libavius in 1616.[20] This short *Apologia* was to be the first of a steady stream of publications he was to write over the next twenty years. In these works he described in great detail a universal system founded upon the macrocosm-microcosm analogy which depended upon the universal life spirit. This was for him the true Chemical Philosophy. I will say in passing that there are a number of aspects of his work that are of interest to all historians of science, but we do not have time to discuss them in detail here. The important point for us is that his mystical interpretation of the Chemical Philosophy became the subject of a heated debate.

The first volume of Fludd's many folios describing the macrocosm and the microcosm appeared in 1617.[21] Here he attacked the Copernican system and he turned to musical harmonies to explain the solar system. The latter for him was the proper application of mathematics to astronomy. This section was noted immediately by Johannes Kepler who delayed the publication of his *Harmonices mundi* (1619) to prepare an appendix directed against Fludd's concept of mathematics.[22] Here and in a second work (1621) Kepler emphasized the sharp distinction to be made between the true mathematician (Kepler) on the one hand and the chemist, the Hermeticist and the Paracelsian (Fludd) on the other. But although this distinction was hardly as clear as Kepler might have wished, it is true that the meaning of mathematics for Kepler was quite different than it was for Fludd. The latter sought mysteries in symbols according to a preconceived belief in a cosmic plan while the former insisted that his hypotheses be founded on quantitative, mathematically demonstrable premises. If an hypothesis could not accommodate his observations, Kepler was willing to alter it; Fludd would not.

Although the Fludd-Kepler exchange is of considerable interest, the scope or the reaction to Fludd's publications among French scholars was to be much broader.[23] In France the Paracelsian medical debates of the late sixteenth and early seventeenth centuries had been accompanied by a steady flow of new and reprinted chemical texts. A new crisis occurred in 1624 when fourteen alchemical theses were defended at the residence of an influential Hermeticist. The meeting was dispersed on the order of the Parlement of Paris, the Doctors of the Sorbonne officially condemned the theses, and, before the end of the year, Jean Baptiste Morin had published a detailed *Réfutation* of the alchemical position.

There may be no more influential figure in early seventeenth century French science than Father Marin Mersenne (1588-1648). A friend of Descartes, Galileo and Gassendi, he kept the savants of Europe up to date on current scientific research through his extensive correspondence. In two works published in 1623 and 1625 it is clear that he felt that a truly mathematically based science must first overcome the claims of the Chemical Philosophers. But although he was deeply concerned about the theological implications of the alchemical position, he did not reject alchemy *in toto.* To avoid wild speculations in the future he sought the establishment of national alchemical academies which could police the field while taking as their goal the improvement of the health of mankind and the reform of science in its entirety. For Mersenne a reformed alchemy would steer clear of religious, philosophical and theological questions. Dreams and speculations such as the chemical interpretation of the Creation must be rejected if the subject was to gain the approval of the Catholic Church.

In the course of his work Mersenne had singled out Fludd as an heretic and a magician. Deeply wounded, Fludd replied in two monographs in which he again described the analogy of the macrocosm and the microcosm, the harmony of the two worlds, the significance of the vital spirit, and its dispersal through the arterial system. True alchemy, Fludd insisted, has as its goal the establishment of the entire Chemical Philosophy as a basis of explanation for both man and the universe.

It is clear that Fludd's understanding of an «alchemia vera» was precisely what Mersenne sought to avoid. Above all, Fludd was disburbed by Mersenne's warning that alchemists should disassociate themselves from religious matters. This subject is one that seeks to comprehend the Creation and the spirit of life. Nature and supernature were surely united — and chemistry serves as a key to both.

In despair Mersenne sought support among other European scholars against Fludd's system of the world. To this end he sent a collection of Fludd's works to his friend, Pierre Gassendi, with an appeal for aid late in 1628. Gassendi, of course, rivals Mersenne as a founder of the mechanical philosophy because of his scholarly refutation of the Aristotelian philosophy and his efforts to establish an atomic explanation of matter. In little more than two months after receiving the books from Mersenne, Gassendi had completed his critique. In an impassioned passage Gassendi complained of Fludd's views which would make «alchemy the sole religion, the Alchemist the sole religious person, and the tyrocinium of Alchemy the sole Cathecism of the Faith.» It is a devastating attack which is made even more interesting because of Gassendi's rejection of William Harvey's views on the circulation of the blood. [24] Mersenne — already aware of the *De motu cordis* had sent this book to Gassendi along with the works of Fludd surely believing its author to be a disciple of his adversary. Gassendi rightly distinguished between the experimental work of the one and the «mystical anatomy» of the other. Fludd had earlier written of a mystical circulation of the spirit of life in the body through the arterial system, but Gassendi rejected both Harvey and Fludd because he insisted on the Galenic system of the blood flow. This was the first significant debate on the Harveyan circulation and it is interesting to see Harvey interpreted as a disciple of Robert Fludd in a work centered on the fallacy of the chemical approach to nature.

Among those contacted by Mersenne in his crusade against Fludd was Jean Baptiste van Helmont (1579-1644) who was to carry on a rich correspondence with the French savant. [25] In one of the earliest letters (1630) van Helmont answered a query on the value of Gassendi's then recent reply to Fludd. The Belgian-physician-chemist was unequivocal as he referred to the «fluctuantem Fluddum» who was a poor physician and a worse alchemist — a superficially learned man on whom Gassendi should not waste his time.

Yet, if van Helmont was at least temporarily included within the Mersenne circle, his work reveals that he had been deeply influenced by that of the earlier Paracelsians. In his first publication (1621) he had praised the macrocosm-microcosm analogy, Paracelsus and his three principles, and he had termed magic «the most profound inbred knowledge of things.» [26] This work was to lead to his prosecution for heresy by the Spanish Inquisition and to his imprisonment followed by house arrest. Among other charges he had been accused of following Paracelsus and his disciples in preaching his Chemical Philosophy, thus having spread more than «Cimmerian darkness» over all the world.» [27]

Van Helmont's later works show a more critical approach, but there remains much to connect him with his chemical predecessors. Throughout we meet a strong plea for reform. It was necessary to «destroy the whole natural Phylosophy of the Antients, and to make new the Doctrines of the Schooles of natural Phylosophy.» [28] Ancient science and medicine was characterized as «mathematical» and logical and this must be avoided at all costs in favor of a truly observational approach to nature. The new philosophy proposed by van Helmont was one that would seek to reject any concept of nature interpreted primarily through mathematics. [29]

Throughout van Helmont's work may be noted the close association of nature and religion. Once again we are told to look first at the account of Creation in *Genesis.* This — as in the case of Fludd — ascertains the order of Creation and the true elements which are water and air. The *tria prima* may be useful to the practicing chemist, but they are not elementary. [30] The key to nature is to be found in fresh observations — and it is chemistry that offers us our greatest opportunity for truth.

> I praise by bountiful God, who hath called me into the Art of the fire, out of other professions. For truly Chymistry, hath its principles not gotten by discourse, but those which are known by nature, and evident by the fire: and it prepares the understanding to pierce the secrets of nature, and

> causeth a further searching out in nature, than all other Sciences being put together: and it pierceth even unto the utmost depths of real truth.[31]

Coupled with this was an awareness that quantification — understood here as laboratory weights and measurements rather than mathematical abstraction — might well offer new insight.[32] Although the willow tree experiment is the best known example, one may also cite his interest in the determination of specific gravities and a more accurate scale of temperatures for laboratory work. In the course of his work he was led to insist on the indestructibility of matter and the permanence of weight in chemical change.

Van Helmont's medicine reflects his background.[33] Unwilling to accept the ancient medical texts, he was also disturbed by those who would accept everything ascribed to Paracelsus. Thus, in these later works, van Helmont rejected a doctrine of the microcosm which postulated man as an exact replica of the greater world.[34] Still, this did not prevent him from calling attention to numerous similarities that were to be found in both man and nature as a whole. As an example, he considered disease in man to be localized phenomena, and although considerably more complex, similar in many ways to the growth of metals in the earth. Nor was van Helmont less concerned than Fludd with the vital spirit.[35] His belief in its existence in the blood was to be an influential factor in his firm rejection of blood letting. No less important was his chemical investigation of digestion which was to lead to the concept of acid-base neutralization which he described with both physiological and inorganic examples.[36]

Like other Chemical Philosophers van Helmont sought educational reform. We have noted this demand before, but van Helmont's work appeared in print at a time when numerous plans for educational reform were being proposed and being seriously considered. Convinced that the study of nature was the only goal to medical progress — and to a knowledge of our Creator — he wrote bitterly of his own education which had led to no certainty and which had caused him at one time to decline a master's degree.[37] If we are to progress, he argued, we must reject the Aristotelian studies of the universities and build a new science upon an entirely new educational system. Students should begin their work with simple subjects: arithmetic, geometry, geography, the customs of nations and an examination of plants and animals. After three years of these elementary studies the young men might then proceed to the important part of their education, the study of nature. But the study of nature for van Helmont may be carried out satisfactorily in only one way, through chemical examination. These studies, carried out for four years, must not be accomplished

> ...by a naked description of discourse, but by a handicraft demonstration of the fire. For truly nature measureth her work by distilling, moystening, drying, calcining, resolving, plainly by the same means, wereby [chemical] glasses do accomplish these same operations. And so the artificer, by changing the operations of nature, obtains the properties and knowledge

> of the same. For however natural a wit, and sharpness of judgement the Philosopher may have, yet he is never admitted to the Root, or radical knowledge of natural things, without the fire. And so everyone is deluded with a thousand thoughts or doubts, the which he unfoldeth not to himself, but by the help of the fire. Therefore I confess, nothing doth more fully bring a man that is greedy of knowing, to the knowledge of all things knowable, than the fire.[38]

Convinced that the new Chemical Philosophy must supersede the now defunct studies of the schools, van Helmont predicated that it would be a small wonder to see just how much a student educated as he suggested «shall ascend above the Phylosophers of the University, and the vain reasoning of the Schooles.»

Van Helmont's call for a chemical reform of education may have been pursued most ardently in England. Noah Biggs demanded a reform of the curriculum of medicine and natural philosophy that was clearly based upon the Helmontian blueprint in 1651 while John French wrote of «a famous University beyond Sea, that was faln into decay» that had been restored to its former glory through the encouragement of chemistry. More important was a debate in 1654 relating to education at Oxford and Cambridge.[39] In that year John Webster, a non-conformist minister, a surgeon, and a chemist, demanded the abolition of scholastic methods at these seats of learning. He argued that divine truths would only be imparted through God's revelation to man in the Bible — and also through the study of His created nature. Displaying a broad knowledge of recent work in medicine and the sciences, he called for an emphasis on new techniques in mathematics, the teaching of the Copernican system and the Harveyan circulation, and the use of atomic explanations. But above all, he felt that it was the Paracelsian and the Helmontian chemists who truly offered an observational and experimental program that might be emulated by others. The essential key must be chemistry for this teaches the unfolding of nature's secrets through the use of manual operations. Like van Helmont, Webster noted that no philosopher is admitted to the root of science without a knowledge of fire. He added that one year of work in a chemical laboratory would prove more beneficial than centuries of disputes over the texts of Aristotle.

For the progress we might demand from this new science, this chemical science, nothing less than the destruction of the scholastic system seemed necessary. And when Seth Ward and John Wilkins rose immediately to challenge Webster on behalf of the current curriculum,[40] they soon found that their own dream of a new science founded on mechanical principles had more in common with the Aristotelians entrenched at the universities than that of the «experimental» chemists. There is no doubt that the chemists were the most vocal proponents of an educational reform based upon a program of the observational and experimental study of nature.

Of course we all know that neither Oxford nor Cambridge — much less the other universities of Europe — reorganized their curricula to accommodate the reforms demanded by the chemists. The «Philosophia nova» that triumphed was that of the mechanists rather than that of the chemists. This may have been due to the superiority of their science — or it may have been due — at least partly — as Jacob has suggested for the English scene, to the rejection of the Chemical Philosophy by latitudinarian churchmen who associated with it the religious enthusiasm and the radical politics that has led to the Civil War.[41] We do know that few alchemists or Helmontian Chemical Philosophers were made fellows of the Royal Society of London at its founding in 1662. Nor do we see them being admitted in strength to the other national scientific societies of the late seventeenth or the eighteenth centuries. We know from their many new publications that they continued to be active, but their failure to be admitted to the academies shows clearly that they were not part of the new scientific establishment.

However, this does not mean that they had been unimportant in the century prior to the founding of the scientific academies in London and Paris. In that period there had been an enormous quantity of literature authored by the Chemical Philosophers — and we have shown that these works were widely read among a closely knit European circle of scholars. Erastus, Kepler, Libavius, Mersenne, Gassendi and many lesser names were fully informed of the publications of the chemists. Their debate with the chemists had touched on a number of key questions related to the establishment of a new science: the value of the ancient philosophers, the role of fresh observations and experiments, the use of mathematics in the interpretation of nature as it relates to reasoning, to the study of motion, and to the laboratory. Harvey's views on the circulation of the blood were first debated in this context, and here, no less than in the Copernican debates, the relation of science to religion was an important subject of concern. These subjects, and others besides must be of interest to all who are concerned with the period of the Scientific Revolution.

The importance of the Chemical Philosophers might be difficult to understand if we limited ourselves only to medicine: the introduction of chemically prepared medicines, the concept of disease, or even the general problems of medical theory. The Paracelsians were not only physicians. They must be understood also as Chemical Philosophers of nature, scholars who consciously sought the replacement of ancient knowledge with a «new philosophy» founded on the twin pillars of true religion and fresh — chemically based — observations. Because of the similarities of the greater and lesser worlds — whether considered literally as with the earlier Paracelsians or metaphorically as with the mature van Helmont — medicine played a central role in their concept of a new knowledge of nature.

It is possible to note the direct influence of these Chemical Philosophers in the general acceptance of the chemical medicines in the course of the seventeenth century, in the persistent chemical search for a vital spirit, and in the widely held chemically oriented systems of physiology proposed by authors such as Willis or de la Böe Sylvius. It is, indeed, difficult to understand Charleton Boyle, Mayow, Becher, Glisson or even Isaac Newton without some knowledge of this then widely accepted approach to nature.[42] Medical topics are frequently discussed and understood in relation to element theory, the chemical theory and properties of the earth and the generation and growth of metals. Not only must the student of this Chemical Philosophy be prepared to investigate these topics, he should also expect to find in the chemical texts other subject as widely separated as educational, economic, and agricultural reform.

There are some who would argue that we may safely ignore these works. I referred in an earlier lecture to Mary Hesse's defence of a more traditional interpretation of the Scientific Revolution. She rejected current research on the Hermetic sources and then went on to state that «throwing more light on a picture may distort what has already been seen.»[43] We may ask in reply whether the picture we have seen in the past has been distorted and suggest that this might be clarified with additional light. Surely the study of the large body of Paracelsian-Helmontian literature of the seventeenth century will add immeasurably to the complexity of our understanding of the rise of modern science. But can we ignore it if this literature was read and debated at the time? To assume that our ignorance of it will ensure a more accurate interpretation seems to me to be a most remarkable conclusion. Traditional accounts should be called what they are — the history of physics and astronomy in the sixteenth and seventeenth centuries.

If we are to act as responsible historians we must try to judge historical events in the context of the period they occurred. If we do this for the Scientific Revolution we find that a major debate centered on the acceptance or the rejection of the Chemical Philosophy. For this reason the debate must play a fundamental role in our future histories of science and medicine as well.

NOTES

1. This lecture was drawn heavily from numerous studies of mine written over the past twenty years. I refer the reader especially to «The Chemical Philosophers: Chemical Medicine from Paracelsus to van Helmont,» *History of Science 12* (1974), 235-59. This was based upon part of the manuscript of *The Chemical Philosophy: Paracelsian Science and Medicine in the Sixteenth and Seventeenth Centuries,* 2 vols. (New York: Science History Publications) which was not published until 1977. My view of the relationship of the Chemical Philosophy to other aspects of the Scientific Revolution was discussed in my *Man and Nature in the Renaissance (Cambridge et al.: Cambridge U.P., 1978).*

2. *See Allen G. Debus, «Mathematics and Nature in the Chemical Texts of the Renaissance,» Ambix 15* (1968), 1-28, 211.

3. On the Aristotelian elements and the Paracelsian principles see Walter Pagel, *Paracelsus: An Introduction to Philosophical Medicine in the Era of the Renaissance* (Basel/New York: Karger, 1958), pp. 82-104.

4. Debus, *Chemical Philosophy 1,* 58-59.

5. *Ibid.,* pp. 69-70.

6. Petrus Severinus, *Idea medicinae philosophicae* (1571; 3rd ed., Hagae Comitis: 1660), p. 39.

7. *Ibid.,* pp. 55-56 and *passim.*

8. *Ibid.,* pp. 57-58 and see also note 3 above.

9. Allen G. Debus, «The Pharmaceutical Revolution of the Renaissance,» *Clio medica 11* (1976), 307-17.

10. Allen G. Debus, «Edward Jorden and the Fermentation of the Metals: An Iatrochemical Study of Terrestrial Phenomena» in Toward a *History of Geology: Proceedings of the New Hampshire Inter-Disciplinary Conference on the History of Geology, September 7-12, 1967,* Cecil J. Schneer, ed. (Cambridge, Mass.: M.I.T. Press, 1969), pp. 100-21.

11. Allen G. Debus, «Palissy, Plat and English Agricultural Chemistry in the 16th and 17th Centuries,» *Archives Internationales d'Histoire des Sciences 21* (1968), 67-88.

12. Debus, *Chemical Philosophy 1,* 96-109.

13. Pagel, *Paracelsus,* pp. 104-12.

14. Studies of particular interest include Robert P. Multhauf, «Medical Chemistry and 'The Paracelsians,'» *Bulletin of the History of Medicine 28* (1954) 101-26 and «The Significance of Distillation in Renaissance Medical Chemistry,» *Bulletin of the History of Medicine 30* (1956), 329-46); Wolfgang Schneider, «Der Wandel des Arzneishatzes im 17. Jahrhundert und Paracelsus,» *Sudhoffs Archiv für Geschichte der Medizin und der Naturwissenschaften 45* 1961), 201-15; *Geschichte der pharmazeutische Chemie* (Weinheim: Verlag Chemie, 1972). See also volumes *1, 5,* and *14* in the series, Veröffentlichungen aus dem Pharmaziegeschichtlichen Seminar der Technischen Universität Braunschweig (Prof. Dr. Wolfgang Schneider): G. Schröder, *Die pharmazeutisch-chemischen Produckte deutscher Apotheken im Zeitalter der Chemiatrie* (Bremen, 1957), H. Wietschoreck, *Die pharmazeutisch-chemischen Produkte deutscher Apotheken im Zeitalter der Nachchemiatrie* (Braunschweig, 1962), *M. Klutz, Die Rezepte in Oswald Crolls Basilica chymica (1609) und ihre Beziehungen zu Paracelsus* (Braunschweig, 1974).

15. The work of Erastus is discussed by Debus, *The Chemical Philosophy 1,* pp. 131-34.

16. For the French debate in the late sixteenth and early seventeenth centuries see *ibid.*, pp. 145-73.

17 The English developments that led to the publication of the *Pharmacopoeia Londinensis* are summarized in *ibid.*, pp. 173-91. Much of the basic research on this publication was done by Georg Urdang. This was summarized by him in his introduction to the reprint of the *Pharmacopeia: Pharmacopeia Londinensis of 1618 Reproduced in Facsimile with a Historical Introduction by George Urdang* (Madison: University of Wisconsin Press, 1944).

18. My views on the work of Robert Fludd and its relationship to the Chemical Philosophy are to be found in numerous articles. Many of these plus additional material are summarized in *The Chemical Philosophy 1,* pp. 205-93.

19. For a discussion of Fludd and the Rosicrucian literature see *ibid.*, pp. 208-24 and *Robert Fludd and His Philosophical Key: Being a Transcription of the manuscript at Trinity College, Cambridge, with an Introduction by Allen G. Debus* (New York: Science History Publications, 1979).

20. Robert Fludd, *Apologia compendiaria fraternitatem de Rosea Cruce suspicionis et infamiae maculis aspersam, veritatis quasi fluctibus abluens et abstergens* (Leiden: Godfrid Basson, 1616) greatly expanded the following year to the *Tractatus apologeticus integritatem societatis de Rosea Cruce defendens* (Leiden: Godfrid Basson, 1617).

21. Robert Fludd, *Utrisque cosmi maioris scilicet et minoris metaphysica, physica atque technica* (Oppenheim: T. De Bry, 1617).

22. Debus, *Chemical Philosophy, 1,* pp. 256-60.

23. The French scene including Fludd's debate with Mersenne and Gassendi are discussed in *ibid.*, pp. 260-79.

24. Allen G. Debus, «Robert Fludd and the Circulation of the Blood,» *Journal of the History of Medicine and Allied Sciences 16* (1961), 374-93; «Harvey and Fludd: The Irrational Factor in the Rational Science of the Seventeenth Century,» *Journal of the History of Biology 3* (1970), 81-105.

25. On van Helmont see Debus, *The Chemical Philosophy 2,* pp. 295-379 and, especially for van Helmont's medical views, Walter Pagel, *Joan Baptista van Helmont: Reformer of science and medicine* (Cambridge et al.: Cambridge U.P., 1982).

26. J. B. van Helmont, *Disputatio de magnetica vulnerum naturali et legitima curatione, contra R. P. Joannem Roberti* (Paris: Victor Leroy, 1621). An important work, this clearly shows a different man than do the later tracts published in the post-1648 opera.

27. On his prosecution and arrest see Pagel, *van Helmont,* pp. 10-13.

28. J. B. van Helmont, *Ortus medicinae. Id est, initia physicae inaudita. Progressus medicinae novus, in morborum ultionem, ad vitam longam* (Amsterdam: Ludovicus Elsevir, 1648; reprinted Brussels: Culture et Civilisation, 1966), p. 6 (from the «Promissa authoris»).

29. The relevant texts are cited in Debus, *Chemical Philosophy 2,* pp. 112-17.

30. *Ibid.*, pp. 317-22.

31. van Helmont, *Ortus,* p. 463 [from the «Pharmacopolium ac dispensatorium modernorum» (sect. 32)]. Here I am citing the contemporary English translation by John Chandler [*Oriatrike or Physick Refined...* (London: Lodowick Loyd, 1662), p. 462].

32. Debus, *Chemical Philosophy 2,* pp. 327-29.

33. The most complete discussion of van Helmont's biological ideas and theory of disease is to be found in Pagel, *Van Helmont,* pp. 96-198.

34. van Helmont, *Ortus,* p. 328.

35. Allen G. Debus, «Chemistry and the Quest for a Material Spirit of Life in the Seventeenth Century» in *IV Colloquio Internazionale del Lessico Intellecttuale Europeo* (Rome, 7-9 January 1983) (in press).

36. Pagel, *Van Helmont,* pp. 129-40.

37. Van Helmont's short autobiography, the «Studia authoris» is printed in all editions of the *Ortus* and the *Opera.*

38. *Van Helmont, Ortus,* pp.49-50 [from the «Physica Aristotelis et Galeni ignara» (sects. 9-11)]. Here I cite the Chandler translation (see above note 31), p. 45.

39. John Webster, *Academiarum Examen, or the Examination of Academies...* (London: Giles Calvert, 1654) reprinted in Allen G. Debus, *Science and Education in the Seventeenth Century: The Webster-Ward Debate* (London: Macdonald and New York: American Elsevier, 1970), pp. 67-192.

40. Reprinted in Debus, *Science and Education, pp. 193-259.*

41 *Margaret D. Jacob, The Newtonians and the English Revolution 1689-1720* (Ithaca: Cornell U.P., 1976).

42. These connections are drawn in greater detail in Debus, *Chemical Philosophy 2,* pp. 447-537.

43. Mary Hesse, «Reasons and Evaluation in the History of Science» in *Changing Perspectives in the History of Science: Essays in Honour of Joseph Needham,* Mikulás Teich and Robert Young eds. (London: Heinemann, 1973), pp. 127-47 (143).

INDEX TO NAMES CITED

北京大学出版社教育出版中心

部分重点图书

五、学术道德与学术规范系列读本(学习之道丛书)

科技论文写作快速入门
给研究生的学术建议
如何撰写与发表社会科学论文：国际刊物指南
学术道德学生读本
做好社会研究的10个关键
阅读、写作与推理：学生指导手册
如何为学术刊物撰稿：写作技能与规范(英文影印版)
如何撰写和发表科技论文(英文影印版)
社会科学研究的基本规则
如何查找文献
如何写好科研项目申请书

六、北京大学研究生学术规范与创新能力建设丛书

学位论文写作与学术规范
传播学定性研究方法
法律实证研究方法
高等教育研究：进展与方法
教育研究方法：实用指南(第五版)
社会研究：问题、方法与过程(第三版)

七、古典教育与通识教育丛书

苏格拉底之道
哈佛通识教育红皮书
全球化时代的大学通识教育
美国大学的通识教育：美国心灵的攀登

八、高等教育与全球化丛书

高等教育变革的国际趋势
高等教育全球化：理论与政策
发展中国家的高等教育：环境变迁与大学的回应

九、北大开放教育文丛

教育究竟是什么？100位思想家论教育
教育：让人成为人——西方大思想家论人文与科学

十、其他好书

科研道德：倡导负责行为
透视美国教育：21位旅美留美博士的体验与思考
大学情感教育读本
大学与学术
大学何为
国立西南联合大学校史(修订版)
建设应用型大学之路
中国大学教育发展史